WHY STUDY GEOGRAPHY?

The *Why Study* Series

Studying any subject at degree level is an investment in the future that involves significant cost. Now more than ever, students and their parents need to weigh up the potential benefits of university courses. That's where the *Why Study* series comes in. This series of books, aimed at students, parents and teachers, explains in practical terms the range and scope of an academic subject at university level and where it can lead in terms of careers or further study. Each book sets out to enthuse the reader about its subject and answer the crucial questions that a college prospectus does not.

Published

Why Study History? — Marcus Collins and Peter N. Stearns
Why Study Mathematics? — Vicky Neale
Why Study Geography? — Alan Parkinson

Forthcoming

Why Study Languages? — Gabrielle Hogan-Brun

WHY STUDY GEOGRAPHY?

BY ALAN PARKINSON

Copyright © 2020 Alan Parkinson

Published by London Publishing Partnership
www.londonpublishingpartnership.co.uk

All Rights Reserved

ISBN: 978-1-913019-15-0 (pbk)

A catalogue record for this book is
available from the British Library

This book has been composed in
Kepler Std

Copy-edited and typeset by
T&T Productions Ltd, London
www.tandtproductions.com

Printed and bound in Great Britain
by Hobbs the Printers Ltd

Cover image

*A geographer won't just see a pile of beans when they look at
the cover of this book. They'll wonder whether the beans were
grown in Cambridgeshire or Kenya. They'll understand that
consumer purchases of Kenyan beans open up opportunities
for the discussion of food miles, carbon footprints and the
environmental impact caused by their cultivation. They'll
think about the inequalities of global trade and the extent
to which both Kenya's economy and the individual farmers
rely on agriculture. They might even find themselves
musing about water security issues, sustainable food
production, plastic pollution, problems of pesticide use…*

CONTENTS

ACKNOWLEDGEMENTS

First and foremost, thanks to Sally, Ella and Sam for their continued patience, and to Carl Lee for putting his PhD first.

The help of the Geographical Association (GA), and my colleagues there, and of the Royal Geographical Society (with IBG) is gratefully acknowledged.

Thanks also to Richard Allaway (Geography all the Way), Steve Brace (Head of Education and Outdoor Learning at the RGS-IBG), Rob Chambers (St Ivo School), Sam Clark (London Publishing Partnership), Professor Danny Dorling (University of Oxford), Professor Corinna Hawkes (Department of Food Policy, City of London University), Professor Ben Hennig (University of Iceland), Joseph Kerski (Education Manager, esri), Alan Kinder (Chief Executive of the Geographical Society), Elaine Owen (Ordnance Survey), Kit Rackley (geography educator), Emma Rawlings Smith (Leicester University) and Professor Joe Smith (Director of the RGS-IBG).

INTRODUCTION

What is this book about and who is it for?

You can travel the seas, poles, and deserts and see nothing. To really understand the world, you need to get under the skin of the people and places. In other words, learn about geography. I can't imagine a subject more relevant in schools. We'd all be lost without it.

— Sir Michael Palin, traveller, writer and actor

IN THIS BOOK I SET out to answer the very important question posed in its title: why study geography? Along the way, I'll also provide answers to several other important and related questions:

- ▸ What **is** geography?

- ▸ **Why** is geography so important?

- ▸ **Where** can you study geography, and which (type of) university course might be best for you?

- ▸ How has the subject we call geography **changed over time**, and why does it matter now more than ever?

- ▸ What important **knowledge and skills** will you gain and develop through a study of geography?

- ▸ What **careers** are available that will allow you to use your geography qualifications to full advantage?

- ▸ What do those who work with geography for their careers actually **do** with the subject knowledge and skills they have acquired?

If you're approaching GCSE or A level options time (or the equivalent exams in your country if you're not in England, Wales or Northern Ireland), this book can help you work out why adding geography to your option choices would be in your best interests. You will be continuing in some form of education, apprenticeship or training in the UK until the age of 18, so use this time effectively. If, on the other hand, you're getting close to finishing your compulsory education, you might be wondering whether to take your geographical studies to university level, or you may feel you're ready to move into the world of work. Whether you decide to apply to do a geography degree or to take the skills you have learned in your geography lessons with you into employment, I hope this book will prove informative.

Or perhaps you're the parent or guardian of a prospective geography student and want to know a bit more about what they could be opting for. They may come to you asking for help to make this important decision: reading this book should help you speak knowledgeably about the subject and the potential benefits of choosing geography. You may even find that you're tempted back into studying the subject yourself: learning is a lifelong process after all.

While writing this book, I spoke to lots of people who have geography-related degrees and to others who didn't study geography to that level but who have used geography in their current or previous careers. I will use what they told me, along with my own experience and research, to help me explain why geography would be such a *powerful* option for you or your child, and to help you appreciate why geography is viewed positively by many employers, from small companies to multinationals. Geographers are currently in high demand, as we shall see. This applies to all countries. Although much of this book focuses on the UK, lots of it is applicable to other national contexts.

Above all else, geographers study the world, particularly the interesting aspects of the world, thereby giving them an understanding of both its complexity and its fragility. Geography is great fun, and it's also addictive: uncovering one aspect of the subject always leads to a new set of questions. The scrolling landscape that passes by while you sit on a train or aircraft is far more interesting than anything you could scroll past on your smartphone screen. Geographers always try to bag the window seat for a reason.

Many potential employers view those who've studied geography as having the ability to see things *synoptically* or *holistically*, which means that they have the ability to make important connections between different topics, seeing 'the bigger picture' – not surprising when you think about the scale of thought that much geographical thinking requires of a person.

Geographers are also problem solvers, with an ability to see solutions as well as identify when problems might occur. Many industries are interested in this way of thinking. Young geographers will develop a wide range of skills during their studies, some of which are subject-specific and some of which are more general. Through their use, geographers are able to critically evaluate data and imagery that you encounter in your life and will help support the development of a curious mindset that will become a lifelong habit. You were curious enough to pick up this book and start reading it: a very good start.

There's another connection that's important here and that is thinking about the future. At a time when you are thinking about future studies or careers, it's worth remembering that geographers generally have a future-oriented approach. They are always thinking about what happens next. The growing area of land change science is concerned with the impact of our activities on the landscape, for example; geomorphologists are likely to look at a cliff and wonder how long it will stay standing for. Geographers always have one eye on the future and on how to make it better.

This book will also explain why you should *definitely continue to opt for geography* at your school, college or academy. It will identify some topics that are commonly part of a geography undergraduate course, it will suggest geographical elements that are well worth studying and discuss where you might study them, and it will showcase some of the exciting career opportunities that geographical qualifications can lead you into. I will provide guidance on how to make the most of your time at university and discuss the investment of time and money that doing so requires.

Steve Brace, Head of Education and Outdoor Learning at the Royal Geographical Society, is clear on the wider value of geography:

Young people are entering an environment of high tuition fees and a competitive job market, so they rightly want to know where geography can take them beyond school and university. For all students considering their next steps, remember that national statistics show undergraduate geographers are more likely than almost any other students to enjoy and complete their degree, and that geographers experience above average rates of graduate employment and earnings (for female geographers, that's up to 10% more than the average).[1]

Geography is a subject that's very much on the up at the moment. The number of students taking GCSE geography in the UK in 2019 was higher than it has been in nearly 20 years, with almost 9,000 more candidates sitting the examination than did so in 2018: that's a 3.5% increase on the previous year and a 50% increase in applicants compared to the equivalent figure for 2011. What is more, this was the eighth successive year in which candidate numbers have increased. Perhaps you were one of those extra students – if so, great choice! Could it now be time to take your studies to the next level.

Almost 270,000 students took geography GCSE in England, Wales and Northern Ireland in 2019, making the subject the sixth most popular at GCSE level. This is even more impressive than it might seem when you consider that the three most popular GCSEs were the compulsory subjects of English, maths and science, and that some schools also require students to take a GCSE in religious studies.

At a briefing meeting in August 2019, Derek Richardson (vice president of the global publishing company Pearson) suggested that one of the reasons for this recent growth in geography's popularity was that the awarding body was seeing an increase in entries across all humanities and social and political subjects:

It's especially interesting, given the impact of climate change and current political events with Brexit and the situation across the world. Students are perhaps choosing subjects that they feel will be most interesting and relevant to them in their futures.[2]

Richardson also suggested that when students were making their option choices, geopolitical events 'were probably more interesting at that time' than they had been for a while. He added that it was noteworthy that students' interest in global events was affecting their choices.

In an era of 'fake news', the criticality that geography provides is more important than ever. Just as history explores the reliability and objectivity of source material, the countless interactions of people in a city create new geographies every day, and the media can misrepresent some of the changes that are happening to suit their agenda, often blaming certain groups for certain events. University-level geography, in particular, will equip you with a general antidote to fake news, particularly when it is combined with the elements of geopolitics and statistical analysis that are likely to form part of university modules.

The number of A level geography entries also increased last year, with around 35,000 students taking the qualification in England, Wales and Northern Ireland. The Royal Geographical Society analyses these statistics each year and has suggested that many of the additional entries are from students who were previously thought less likely to study geography: a sign that the *value* of the subject is increasingly being recognized, including by some less academically inclined students who buy in to the practical and relevant nature of geography and appreciate what the subject has to offer them.

One thing is clear: all the greatest challenges we currently face come under geography's remit. As the writing of this book was being completed, the greatest challenge the world has faced for many years emerged with alarming pace, changing life for everyone. The much greater challenge of climate change is still there, in the background, playing the long game, but the emergence of Coronavirus Disease 2019 illustrated the importance of geography, both as a subject and as a reality of daily existence. Where people were located made a dramatic difference to their experience, but, equally, we were all connected by the same threat. The Covid-19 pandemic is fundamentally geographical in nature.

Our globalized world means that more than half a million people could be in the air, taking commercial flights, at any point in time. During the period when the virus began to emerge, international flights continued to transfer people away from the Chinese province of Hubei, where the first outbreaks emerged. The way the disease initially stays undetected meant that people carried the virus around the world within twenty-four hours. Arriving at their destination, they dispersed from the airport onto crowded public transport, queued in coffee shops and attended sporting and cultural events. As case numbers began to grow, geospatial organizations started to use their skills and resources to help with the coordination of the response and the tracking of cases.

The all-encompassing global impact of Covid-19 triggered an immediate hiatus in many of the systems that had become part of our daily lives, and that formed an important focus of the geography curriculum studied in schools and colleges. We have become used to global supply chains working efficiently, to commuting to school or work on packed trains, to cheap foreign travel, to plentiful culture, and to easy access to food and entertainment. Our lives are connected, and those connections needed to be temporarily broken and then slowly reinstated.

In a piece in *Wired* in April 2020, David Wolman explained the sudden importance to many of a key geographical word: 'where'. Technology – and particularly the internet – has blurred distance for us, making the far away seem very close, but geography never went away. Wolman defines this very simply: 'I'm here and you're there – boom, geography'. Where you were suddenly mattered more than ever.

The pandemic has made us change how we view space. Visiting a supermarket during the time of lockdown turned into a dance as people navigated the aisles, mindful of where other people were. At the same time, the very presence or absence of food items on the shelves was either a sign of local, national and global supply chains still working or of those linkages facing short-term stresses. Other places aren't disconnected from us: that's Wolman's message. Powerful stuff![3]

This should also remind us of Waldo Tobler's 'First Law of Geography' (1970):

> Everything is related to everything else, but near things are more related than distant things.

Proximity has suddenly gained greater significance.

Researchers in geography departments are incredibly active, exploring many of the impacts of the climate emergency on the planet and involving themselves in the search for solutions. Understanding the causes of changes in a natural system is an important area of focus for geographical study. The production of maps is a vital part of this, often using geographical information systems (GIS) to assess patterns and changes over time. Many GIS maps, including the now-famous Johns Hopkins University dashboard, tracked the growth in cases of Covid-19 as it spread from its original source. This work is not only important, but exciting to be involved in as well. It also tells us that geography in the US, as in the UK and other countries, is often paired with other subjects. At Johns Hopkins, for example, geography is paired with environmental engineering.[4]

Geography involves helping us to imagine and shape what the future might bring. It is no more about knowing the capital cities of countries than history is about memorizing the dates of important events. Geography provides a way of seeing the world with what the Geographical Association has previously called 'a different view': a lens through which the world can be seen in a particular way. You may not remember a time before the existence of Google Earth, which allows you to visit any place on the earth's surface virtually, or before the arrival of the smartphone, which provides access to the world's knowledge (or can make you look like a dog with a Snapchat filter). These tools, and others, are used (slightly more seriously) by geographers to help them make sense of the patterns and processes that are currently reshaping the world. Geography also brings its unique exploration of context and place, helping those who study it to appreciate the complexity of the world. Studying geography can help you be part of that important work. Are you ready for that?

This book has been written specifically to give you an overview of what's currently happening in the world of geography, including information about the important and exciting opportunities that are waiting for you in universities, technology (and other) companies, the financial sector, the geospatial industry and international organizations. It will explain *why* geography is a subject that deserves to be studied, that will repay your involvement with improved job prospects and degree satisfaction, and that will create a new world of opportunity. I've spoken to lots of people who have used their geography qualifications and skills in a wide variety of ways, and regardless of their day job, they understand that seeing the world as a geographer offers them a very special perspective.

At this point, we could mention some celebrities who studied geography to degree level. Common names on such lists include Prince William, who made the wise decision to switch to a geography degree from the history of art while studying at the University of St Andrews; basketball legend Michael Jordan; comedian Hugh Dennis; *Countryfile* presenter Ellie Harrison; and Matthew Pinsent and James Cracknell, who won multiple Olympic gold medals in rowing. The former prime minister Theresa May is a geography graduate too. Many TV weather presenters, including *ITV Weather*'s Lucy Verasamy, also have a geography background, as the complex nature of the atmosphere is often studied during geography degrees, alongside elements of physics and mathematics.

However, it seems very unlikely that you would base such an important decision about the next stage of your life on whether this or that celebrity studied a particular subject. And anyway, while such lists can be interesting, most subject disciplines could produce something similar, so they don't end up telling us much.

I'll explain the questions you should ask before choosing a university in the UK (many of which would be relevant questions to ask whichever country you choose to study in), I'll advise you on how to thrive on your chosen course, and I'll outline some of the areas in which geospatial thinking touches all of our lives, wherever we live. And the book will finish by providing a toolkit of ideas, resources and reading that you can engage with to help develop your own inner geographer. This will include advice from authors, artists, scientists and others who have taken time to focus on their own worldview.

As with the companion volumes in this series, I'm going to fly the flag for my own particular subject, but geographers also appreciate the important connections that we have with other curriculum areas. For Alastair Bonnett, a geography professor at Newcastle

University, geography helps us to apply order to the world in which we live:

> *Geography surrounds us. We walk, drive and fly over and through it. At the same time, geography is distant: geography is the landscape beyond the horizon and the intriguing distance between here and there.*[5]

There's a whole world out there, and geography can help people not only to understand it, but also to thrive in it. We are living in what Stanford ecologist Hal Mooney has called 'the era of the geographer': a time when the discipline is becoming increasingly central to society and to scientific thought. To come back to a definition of geography, we could use the one suggested by Richard Hartshorne in 1959:

> *Geography is concerned to provide accurate, orderly and rational description and interpretation of the variable character of the earth's surface.*

Or perhaps we should turn to the words of Lionel William Lyde, a lecturer at University College London (UCL) about a century ago. He used to arrive at lectures with his notes in his top hat, which is not something you are likely to see now. He described geography as being 'a body of knowledge coupled with an attitude of mind'.

Authors such as Tim Marshall have shown that a clear presentation of certain aspects of the subject can prove very popular with a general readership. His 2015 book *Prisoners of Geography* draws together various strands of the subject to explain a number of geopolitical situations, such as the Turkish–Syrian border and other controversial borders, suggesting that geography will always result in disagreements over place and where borders lie. The book draws on ideas that geographer Sir Halford Mackinder first outlined over a century ago, shaped by his personal experiences across the world. Rivers, he reminds us, act as natural borders and divide land up into territories; mountains can protect, but they can also create a barrier to trade and movement. *Prisoners of Geography* has sold more than 250,000 copies in the UK alone, and the follow up books *Divided* and *Worth Dying For* have proved equally popular. In an article co-written with Sophie Donovan in *Geographical* (the magazine of the Royal Geographical Society), Marshall tells of the terrific response he has had from those studying GCSEs and A levels. He says of the book that he

> *hopes it gives [readers] a geographic or historic context for what they're seeing in front of them, and that it will lead to many different paths... If it's opened a door in your mind, or if it leads to a path that grabs you, that's one of my definitions of success.*[6]

Some geographers have noted the connection with an idea called 'environmental determinism', which has its critics. Geographers should be critical of what they read, as there are many different stories that can be told, and different perspectives are important. This critical thinking about sources is just as important to geographers as it is to historians; indeed, there is an important subdiscipline called historical geography that aims to understand the past.

At the end of each section of the book, I'll suggest some simple things that students could do next to help them decide on their best course of action, and to move their thinking on. These will be practical things, sometimes involving talking to other people, sometimes looking at websites or other materials. I hope this will encourage reflection on what has just been read.

Some things to think about before reading the next chapter

► Talk to your geography teacher or lecturer, if you're in touch with them, and ask them why they chose to study geography. What keeps them interested in the subject after all these years? They may also be able to give you an honest opinion about your ability in geography and to share their ideas about which universities are particularly good for geography.

► Look at your smartphone and think about which apps that you have installed are 'geographical' in nature or make use of geographical data. How many are collecting data about your movements and daily habits as you use them? You might be surprised! I'll explore the role of geolocated data more in a later chapter, but this became significant during the coronavirus outbreak with 'track and trace' systems.

► Which countries and regions of the world have you visited in person? Where are the blanks in your personal experiences? How did visiting the areas you have visited change your views on them? Which world map did you choose, and how does the particular 'projection' used to make the map change how you view the world? Where would you like to travel next when circumstances permit?

CHAPTER 1

What is geography?

There are so many ways in which geography is fundamental to our lives, from physical landscapes, natural disasters, the characteristics of where we live, the energy we use, services we enjoy, the travel we undertake and the networks in which we engage. Geography is a cornerstone in the continuing education of everyone, both young and old, helping to make us more effective local and global citizens.

— Gill Miller, President of the Geographical Association, 2019–20

THIS CHAPTER WILL DISCUSS HOW geography developed as a school subject, what studying geography involves, and why it's so important for everyone to have an appreciation of the power of geography and to be able to think geographically.

So what is geography?

One challenge for geographers when they are asked to describe their subject and what it involves is its *scale*. It is the 'world discipline', with the whole world as its object of study. This is quite a challenge for both teachers and students: where do you start with such a breadth of possible subject matter? It certainly makes it hard to narrow down exactly what it is that geographers *do*. A global scale allows for a huge variety of geographies to emerge. Very few people actually have 'geographer' as their occupation, and you won't see many job advertisements recruiting them. The same issue doesn't apply to some other academic subjects: you will often see someone in the media described as a 'historian', for instance. Fortunately, this diversity can be turned into one of geography's strengths: what many people do in their job *is* geography, even if it isn't necessarily called that. Many TV science programmes are geography in everything but name. The 2020 Netflix series *History 101* is as much geographical in its subject matter as it is historical.

The *Encyclopaedia Britannica* definition of geography, written by the late Professor Ron Johnston, says that the subject involves 'the study of the diverse environments, places, and spaces of Earth's surface and their interactions'. More pithily, he said that geography studies the earth as 'the home of mankind': a good place to start, for sure, although he would perhaps have said humankind today.[7]

Geography is the subject that studies the world in the most sophisticated way, going back all the way to the Ancient Greeks, who were the first to map the world and ponder its mysteries. The Greek scholar Eratosthenes is credited as being the person who gave us the word 'geography', about 2,200 years ago, with 'geo' meaning the world and 'graphy' broadly meaning to write about, suggesting that geography is the science of 'earth description' or 'earth writing'. There is no shortage of stories we could write about the world, and as geographers often remind us, *who* is writing the stories can be significant. We can all carry out some geographical thinking if we consider a place that we have (or haven't) visited and write about it. Even a postcard from a holiday destination (or, more likely, a photo from our social media updates) may include some geographical description of the location, the weather conditions or the local culture.

Other early geographers include explorers such as Marco Polo, Vasco da Gama, Mungo Park, Isabella Bird, James Cook, Freya Stark and even Doctor Livingstone (I presume?). They explored the world from a Western perspective and reported back to people who would never have the chance to follow in their footsteps.

Geographers should therefore be interested in both telling and consuming compelling stories about the world, including grand tales of Polar exploration and scientific discovery, but also 'smaller' stories, such as those of each individual migrant crossing the Mediterranean looking for a new life in the European Union, or of a firefighter in Australia confronting a wildfire that is burning out of control and having to assess whether the wind will change dir-ection and put them at risk. Each story has a context – it is set in a place – and a narrative that connects it with a number of external factors, such as weather and climate, urban or rural landscapes, and the time of year. Marvin Mikesell coined the pithy phrase that geography is 'the why of where'.

Those early geographers were often travellers who came home with stories and other information (sometimes less than truthful) about the world they had seen. They were often the first people from their country to visit these places, at a time when 'terra incognita' (unexplored lands) still existed and sea serpents and dragons were drawn on the maps. We've gone beyond that now, although in 2020 our days of carefree foreign travel have been temporarily curtailed. One response to that limitation for many has been virtual travelling, and a desire to see more of the world – there is a lot of it to see.

The Roman astronomer and mathematician Ptolemy also, very importantly, started to **ask questions about the world**. Why are certain plants found in certain parts of the world, for example? Why do rivers only flood at certain times? Ptolemy himself actually came up with a useful early definition of geography as long ago as the second century CE:

A representation in pictures of the whole known world together with the phenomena which are contained therein.

This idea of images, then, is also really important to geographers – we love pictures. Information on the relative location of places was later used to make maps. Those that have survived are not necessarily small maps of local areas but larger maps showing whole territories, expanding outwards to the 'edges' of the known world as the globe began to be filled in. Although not all geographers are cartographers, all cartographers are geographers.

Maps were also sources of power, as they were often created to signify ownership of territories and resources. They tended to be made by and for people who had both an education and the money to pay for their creation. While maps are still one of the main tools of the geographer, they have arguably changed more than any other with the arrival of new technologies such as geographical information systems; we shall return to these later, as they form an important part of many undergraduate geography courses. These technologies, and collaborative projects such as Open Street Map, have democratized the ownership and production of maps: they are made by and for everyone, adults and young people alike. They have also been joined by the infographic: a combination of an image and text, often beautifully designed, with which you will be familiar from news reports and newspapers. The ultimate expression of maps for many are those found on smartphones: maps that build themselves around you as you move, providing you with the specific information you need to find a tube station, navigate to your workplace or gym, share your location with friends on WhatsApp, or 'stalk' a family member on Snapchat.

If you love maps, then you are a good way towards loving geography. Think of some of the story books you may have read when you were younger. Many had fantastic maps that stirred the imagination of their readers while also giving them a point of reference for the action in the Hundred Acre Wood, Middle Earth, Narnia, Panem or Earthsea.

Geography remains a difficult subject to pin down to specific topics of study. Not all geographers study the same things. It may be more accurate to say that there are **many geographies**, as each of you will have studied slightly different elements at school, depending on the particular interests of your teachers and the decade during which you went there. The specification of geography varies from country to country, too, as does the relative status of geography itself. Likewise, the course content of geography at university will vary depending on the research interests of your lecturers and on your own choice of dissertation topics, as we shall see in a later chapter.

All geographers and geography teachers have their own particular favourites, just as historians may prefer to research the Vikings and their exploits, say, rather than the history of medicine. Nobody can really know all the details of the whole world, so **choices** have been made over the decades as to which particular knowledge is important for students to learn and/or powerful to have when it comes to making decisions. The extent to which students acquire and are able to demonstrate that they can apply this selection of knowledge lies behind all external qualifications. The nature of success in these qualifications in turn dictates the likelihood of success of those applying for university places and particular jobs.

As Noel Castree, a professor of geography at the University of Manchester, suggested in his book *Nature* in 2005:

> *There is no one 'correct' set of things that students should know; there is no one 'proper' way of learning; there are no 'self-evident goals' of education. Instead there are only ever choices about what to teach, how to teach and to what ends... When these choices are made and accepted by a sufficient number of teachers then they tend to become 'common sense'.*[8]

To illustrate the amazing breadth of geography's reach and relevance, here are just a few of the topics that current students may have already studied in recent school geography lessons.

► Geopolitical tension in the Arctic, with Russia presenting a claim to the North Pole to the United Nations in 2016 and President Trump offering to buy Greenland in 2019.

► The rapid melting of glaciers in mountain regions, creating water-supply problems for those communities that rely on them and releasing toxic metals as the glaciers retreat.

► Concerns over the clearance of tropical rainforests to allow soya beans and oil palms to be grown and exported to the UK and elsewhere for animal feed. This process accelerated when the world's attention was elsewhere.

▶ The terrifying scale of bushfires in the 2019–20 'bushfire season' in Australia.

▶ The rapid changes taking place in urban areas such as the King's Cross area of London.

▶ The impacts of tourism in countries such as Iceland, where tourists outnumber the population and put pressure on significant locations on the Golden Circle tour route.

▶ The changing nature of travel and commuting as a result of Covid-19.

I could go on, but you get the picture. Teachers choose certain topics to teach, hoping that students, in turn, find them interesting, giving them an incentive to make their own choice: the choice to continue studying geography beyond the point where they have the option to stop.

Geographers have always explored **the relationship between people and the environment**. This means they are both physical scientists and social scientists. Geographers are able to explain complex physical systems such as the atmospheric system, plate tectonics, ecosystems and the carbon and water cycles, but as social scientists they are also interested in population, trade, conflict, globalization, the growth of cities and world development. We really are very curious about most of the things that matter.

Geography is a key subject in an area called the humanities. Humanities subjects extend across the social sciences and draw on the strengths of each of the other humanities areas, including history to an extent. They are academic disciplines that study aspects of human society and culture. Some university courses are also called 'Humanities', and many universities offer degrees in this area or may locate their geography department within a Humanities Faculty. Geography spans both natural and social science, placing the discipline in a unique position to address global issues.

The German scientist Alexander von Humboldt advanced the subject of geography in the early nineteenth century, exploring how plants were influenced by climate and landscape: for example, he produced early examples of what we now call infographics, showing how the spatial distribution of plants changed with altitude in mountainous areas. This scientific approach became popular. Over time, the subject has evolved as the world has become increasingly complex. New subdivisions appear to reflect new areas of study. For Alexander Murphy, author of a book called *Geography: Why It Matters* (2018), the geography that is studied today 'offers a critically important window into the diverse nature and character of the planet that serves as humanity's home'.[9]

As Carl Lee and Professor Danny Dorling say in their 2016 book *Geography*, the subject is vital for all young people to study. They are clear that if one wants to understand the world in which one lives, one needs geography.[10]

Professor Graham Butt has also said:

With its particular conceptualisations of places and spaces, employment, leisure, citizenship, domestic life, sustainability, interdependence, identity and consumption (to list but a few!), geography and geographers can help to provide young people with a passport to their lives.[11]

This phrase – 'a passport to their lives' – is an important one. Geography as a subject to study is an enabler that allows access to different places (sometimes literally). One key part of modern geographical study is to help you 'think geographically'. Those of you already studying geography will be familiar with this, particularly at university level.

For those currently studying geography in UK schools, the most recent Geography National Curriculum document – which lists what the government thinks students aged between 11 and 14 should be studying in geography lessons – influences what students are taught. It was last updated in 2013, so it's already a little out of date. The 'Purpose of Study' statement, which comes at the start of the document, says the following:

> *A high-quality geography education should inspire in pupils a curiosity and fascination about the world and its people that will remain with them for the rest of their lives. Teaching should equip pupils with knowledge about diverse places, people, resources and environments, together with a deep understanding of the Earth's key physical and human processes.*
>
> *As pupils progress, their growing knowledge about the world helps them to deepen their understanding of the interaction between physical and human processes, and of the formation of landscapes and environments. Geographical knowledge provides the tools and approaches that explain how the Earth's features at different scales are shaped, interconnected and change over time.*

The words curiosity and fascination are important, but we are also reminded that knowledge about these other places is significant. Think back to your own geographical studies at that age, whenever that may have been. Can you remember learning all of those things, or remember which countries you studied in detail? Which elements of your geography lessons did you most like, and which were you less enamoured with? Also bear in mind the different ways you 'see' a place (e.g. your school) compared with your parents. As Sarah James has remarked:

Even when children share the same settings as adults – such as the home or public space, parks and shopping centres – what they expect and what they are expected to do there is likely to differ, and thus we see variations in ways in which children and adults experience the same environment. For example, in parks the children use the space for play, physical and emotional exploration and development of various kinds; whilst for the adults who accompany the children the space may perform a social function, a place to meet and talk to parents and child-minders.[12]

Five key elements of geography

Between 2006 and 2011, just before the current National Curriculum was written, school geography in the UK underwent big changes. Today's geography is much more forward thinking and relevant than it used to be – if you're still in school, ask your parents to tell you what they remember from their own school geography lessons. In 2006, the government provided millions of pounds to fund a project called 'The Action Plan for Geography'. The Geographical Association and the Royal Geographical Society teamed up to support teachers and introduce new geographical thinking. At the time, it was thought that there were five key elements of geography that it was particularly important to introduce young people to during their formal education: 'physical and human connectedness', 'place', 'scale', 'process' and 'skills'. We reflect on each of these elements below.

Physical and human connectedness

Geography is a subject that doesn't separate natural and human-made environments. Humans lead social, economic and cultural lives, but each one of us ultimately lives within a physical environment. This environment provides both opportunities and constraints, and human actions in turn influence that environmental context. Think about topics in which you have studied the problems that living in particular places has caused for people: the 'shanty towns' in cities in South America or central Africa, for example.

Place

Geography is the subject that studies how the environmental, social and economic interactions mentioned earlier actually play out: things like the provision of living space and work, and the travel that takes place between them; or how environmental processes impact on development. Places, and the people who live in them, are in some respects comparable wherever they are on the globe, but there are also ways in which they are unique. Geographers understand the importance of diversity and the causes of inequality, and they realize the importance of location in influencing these factors.

This is a big element. When we say 'place', we don't always mean 'places', such as London or Dubai or the North Pole. We attach meaning to particular locations. The landscape artist Alan Gussow described this beautifully (he wasn't a 'geographer', but we will claim him as one):

> *The environment sustains our bodies. But as humans we also require support for our spirits, and this is what certain kinds of places provide... A place is a piece of the whole environment that has been claimed by feelings. Viewed simply as a life-support system, the earth is an environment. Viewed as a resource that sustains our humanity, the earth is a collection of places.*[13]

That's a powerful thing to say. It means that as geographers we might be more interested in looking at the environment – a unit on rainforests might be just our sort of thing – or we might prefer to look at people and their links to places. This explains why some people's geography dissertations are about topics that those who don't understand the subject might think isn't geography. Geography's breadth, which we referred to at the start of the chapter, is therefore a strength as well.

The idea of a sense of place is crucial to our own experience of living in the world. I'll return to this later, but for now we'll move on to the third element, which continues this global theme.

Scale

Geographers are particularly conscious of the significance of scale, whether it be local, regional, national, international or global. The choice of the scale of study will influence the kinds of question that are appropriate, but geographical study also seeks to connect those different scales. We all exist both locally and globally in our consumption of energy and other resources, and in our impact on physical systems such as climate. Geographers use the term *interdependence* here: everyone relies on everyone else, and no country can stand alone. This may have formed part of your course.

Basically, we are all dependent on each other to some degree. No place is completely isolated. Try existing without water for a few hours and you realize the importance of the hydrological cycle, of hidden infrastructure, of public health and government funding and the numerous other people working to keep the supply reliable and potable.

Those who voted for Brexit didn't necessarily understand that the UK can't separate itself from the rest of Europe. Tectonically speaking, both the UK and Continental Europe sit on the Eurasian Plate – not something that is likely to change any time soon, except in some far-fetched disaster movie. A DEFRA report from 2019 also indicated that the UK is only 64% self-sufficient in food production and will therefore continue to require food imports, many of which currently come from the EU. Wherever you live, your home country relies on others for a great deal, and those countries rely on others in turn. The nature of that reliance is often unequal, and this leads to inequalities between countries at different levels of economic development.

The fourth element is particularly interesting.

Process

Geography is dynamic. Environments, societies, landscapes and places – and their interactions – are constantly evolving over time. Geography is the subject that helps people to understand **why and how** that evolution takes place through the study of environmental, economic and social processes and their effects on shaping places and landscapes. A study of process is an essential part of understanding the reasons for, and the nature of, change.

For me, this is a particularly exciting aspect of geography: the subject never stands still. My main job is being a geography teacher, and that means I need to keep up to date through reading and research. Anyone who studies geography will know that it keeps throwing up new ideas and exciting avenues to explore. The processes we study operate over different timescales. A cliff can stand for thousands of years but be slowly hollowed out and shaped by the action of waves, until one stormy night it collapses and the coastline retreats. Rainforests that took millions of years to develop can be quickly destroyed by fire or chainsaws. Cities grow over time, and this can be tracked by satellite images. A time-lapse film of Dubai's growth would show desert, just a couple of decades ago, being transformed into the intricate modern city of today: a city that is reliant on air conditioning and desalination of saltwater to keep the fairways of its golf courses a vivid green against the surrounding sands.

This brings us to the final and fifth element.

Skills

Classroom and fieldwork studies in geography lead to the development of a wide range of skills. These can be applied to the investigation and understanding of the real world. These skills are much sought after by employers. Geographers develop skills in observation and in posing interesting questions, and they develop important research skills. They gain experience in collecting and analysing data, and they use their communication skills to present, question and discuss findings. Through decision-making exercises, pupils will also begin to predict or suggest alternative scenarios, or **futures**, and to think creatively. This futures thinking is one of the key benefits of thinking geographically, and one you will recognize from your current studies or from memories of school or college.

Information technology plays a key role in geography classrooms, where it may go by the name 'EdTech' when it is used by teachers and students. Geography teachers were early adopters of technology, including Google Earth and similar visualizations, increasingly powerful Geographical Information Systems, and virtual reality. Fieldwork helps to develop team skills as well as specific skills in mapping and analysis.

Another way to think of this is that having skills gives you the ability to do things. I'll come back to what are called GeoCapabilities in Chapter 4. The idea that studying geography provides people with the capability to do different things is significant.

Reflect on these five elements the next time you read a newspaper.

The importance of maps

One crucial tool for geographers is the map. There is an important and ongoing relationship between maps and power. In the present day, it is companies such as Google and Apple that determine how places are represented and what information we are presented with.

Professor Alastair Bonnett has written books called *Off the Map* and *Beyond the Map*. He talks of geography as being one of humanity's big ideas, and at a 2019 meeting of teachers organized by the Prince's Teaching Institute, he said that:

Geography is a kind of evolutionary disposition – people are born with a geographical orientation.

Doreen Massey, who sadly passed away in 2016, was a leading human geographer, exploring notions of place and power, and the changing nature of cities. She wrote widely on the importance of places, influencing how people viewed the concept. Kilburn High Road in London was the scene of one of her famous pieces of writing. She provided her perspective on the importance of the local, and on its connection with the global:

We live local versions of the world and in so doing we have to locate ourselves within a wider global context. We only understand the changes taking place in our own backyard when we begin to understand how changes taking place elsewhere affect our world.[14]

Many geographers since have referred to her work, which provides new perspectives on our everyday lived experiences. She also referred to maps as a surface on which endless stories combine simultaneously.

People associate geographers with maps, and maps are of course one of the tools that we use: they could be said to be our main tool. The UK is fortunate to have the Ordnance Survey (OS), a mapping agency that produces high-quality mapping that is the envy of those in other countries. The classic 1:50,000 Landranger and 1:25,000 Explorer maps provide detailed mapping of both landscape and human features. There's also a whole range of career opportunities at the OS.

Mapping data are used in other ways, by other organizations. A topical example, as I write this chapter in mid 2020, is the use of OS data by American geospatial firm esri (which stands for the Environmental Systems Research Institute) to produce a map of pavement widths in major UK cities. This is part of an exploration of how easily city centre managers would be able to facilitate social distancing as city centres reopened after Covid-19 related lockdown.

One online tool that appeared in 2005 changed the way that many of us view the world. Google Earth isn't really a map at all. It is a huge mosaic of images that changes as one zooms in to reveal more detail (this process is called granularity and happens when you zoom into a map on your phone as well – it needs to happen quickly to be smoothly animated). Google Earth allows us to fly into locations, to see the shapes of the landscape and go down to street level, even inside buildings. Google's cameras have travelled the world, to the most remote locations, to capture imagery. The latest iteration of Google Earth allows for the creation of tours and storytelling, and it is used both in some university studies and for the production of non-exam assessment projects by A level students.

Geography can help avoid some unfortunate embarrassments over location. You may have heard stories of people who misread their satellite navigation devices and ended up in the wrong place: somewhere with a similar spelling, perhaps. Plenty of examples can be found in the news, with lorries getting stuck between buildings on narrow junctions or cars driving into streams or endlessly circling the M25.

Connections between school and university geography

Because they developed separately, there is sometimes a disconnect between school and university geography. When the A levels that some of you will be taking, or have already taken, were planned, an advisory board of academic geographers was involved. They were keen to add certain topics into those courses so that when students arrived at university they were better prepared for undergraduate studies. School geography was developed by academic geographers in the late nineteenth and early twentieth centuries, so the connections have always been there. The last revision of UK examinations saw the appearance of some relatively 'new' topics, which involved some upskilling on the part of teachers. These new topics included

- water and carbon cycles,
- changing places,
- rebranding places and
- drylands.

If you apply for a geography degree and are called for interview (see Chapter 5 for some specific advice on this), you will probably be asked to talk about your ideas of what the subject is, so be prepared to explain your own thoughts about what makes the subject so important. If you asked 20 geographers to answer that question, you would probably get 20 different answers, so work out a response that you are happy with and, most importantly, that you can remember without referring to notes. You may already be studying geography, and these questions may trigger some thoughts about your future employment.

In 2009 the Geographical Association (the subject association for geography teachers and geography educators) commissioned the polling organization MORI to ask a sample of young people their views on geography.[15] Some key findings from the MORI report at the time were as follows.

► Young people saw crime and antisocial behaviour as the most important issue affecting either their local area or the world generally, followed by the economy and jobs. War and terrorism, poverty and hunger, and the environment and climate change were also seen as the global issues that they felt were more important than others.

- ▶ Geography was the subject in which young people had most often learned about/discussed these issues at school, and the one in which they most commonly expected to do so. This shows the relevance of geography to everyday life.

- ▶ The majority of students thought it was important to learn about issues that affect different parts of the world, particularly how the world they live in might change.

- ▶ Most young people thought that not enough time was spent learning about the wider world in school.

A decade on, how much do you think has changed? Recent events have added a layer of complexity to this for many young people, and geography may be even more important in their lives now. If anything, geographical ideas have recently become more significant in studying the issues that affect young people's lives. As Danny Dorling said in an interview in *Geographical*, it is the best option we have for coming up with solutions to our global problems.

New geographies are emerging all the time. This book is being published at a very interesting time for geography, and we will reflect on some of those changes as we continue our story.

Thinking on a global scale

A final element of geographical thinking is that of **scale**. Some geographers spend time peering through microscopes at grains of sand or considering the movements of air currents around individual plants. Other geographers look at the bigger picture.

On Christmas Eve 1968, NASA astronaut Bill Anders was one of the crew members, with Frank Borman and Jim Lovell, on the Apollo 8 mission travelling round the 'dark side' of the moon. The spaceship had already made the same trip three times before; this time, Anders was ready with his Hasselblad camera and took some pictures as earth came back into view. The previous day, Jim Lovell had broadcast grainy black and white images of earth live to those following the NASA mission back home. The mission had achieved a number of significant geographical firsts. The three men became the first humans to travel beyond low earth orbit, to see earth as a whole planet, to enter the gravity of another celestial body (the moon) and orbit it, to see the far side of the moon with their own eyes, to witness the earth rising, and to escape the gravity of another celestial body and re-enter the gravitational pull of earth.

It was some time before the crew safely returned to earth and the film was developed. On the day after Anders took his famous image, the *New York Times* featured a piece by the poet Archibald MacLeish. In it, he wrote:

> *To see the Earth as it truly is, small and blue and beautiful in that eternal silence where it floats, is to see ourselves as riders on the Earth together, brothers on that bright loveliness in the eternal cold.*[16]

The famous image Bill Anders took is known as 'Earthrise'. It triggered a growth in environmental awareness as many people came to understand that earth is our one home in space, and we must protect it. The image has been reproduced many times since and is significant in that it captures the moment when humans first saw their home appearing from behind the shadow of another celestial body and saw the whole of geography for the first time.

A closer look, though, reveals that the original image Bill Anders took was rotated and cropped to achieve the desired effect. Our perceptions of places can be manipulated by the images that are selected for us to see. In a TED talk that has been viewed more than 20 million times, Chimamanda Ngozi Adichie reminds us of 'The Danger of a Single Story', and the need to use a number of sources and question their veracity. Even Tim Marshall's best-selling books generally contain only one opinion (the author's), and the contexts he writes about often require more than one viewpoint to understand them fully. There is seldom only one correct answer, and there will always be quite a few incorrect ones.[17]

The more you read, the better your understanding of geography will be (particularly if what you're reading is the rest of this book). The more stories you immerse yourself in, the more you will be able to think geographically. The following quote from Robert MacFarlane's book *Landmarks* is useful here:

Before you become a writer you must first become a reader. Every hour spent reading is an hour spent learning to write.

If you are going to do geography (earth writing) well, you need to read as many of its stories as you can, whether that is in person or through the words and images of others. You'll already have encountered some of these stories in your existing studies.

I'll finish this chapter with another quote from Danny Dorling, who was even clearer on the importance of geography when he was interviewed for an article in *Geographical* in April 2020:

For the new generation of young people who are passionate about the planet, the degree choice is obvious: geography... You're really allowed to do almost anything you like as a geographer – by combining this new enthusiasm with free thought, geography can truly push boundaries.

I like the idea of doing almost anything you like, and that really is the case with geography degrees. There are so many opportunities in the study of geography, and it doesn't close any doors in terms of careers. Geography can take you all sorts of places, as we shall see in later chapters.

Some things to think about before reading the next chapter

▶ Load up Google Earth (https://earth.google.com/web/) on the Chrome web browser. It's described by Google as 'the Earth's most detailed globe'. Click the Voyager icon on the left (it looks like a ship's wheel) to see a whole host of curated stories or just (re)visit and explore your own local area, or zoom out as far as you can and recreate 'Earthrise' on your own computer or tablet screen.

▶ Which global issues are you most personally concerned about? What practical actions could you take – or have you already taken – to improve your personal response to these issues? How can you communicate what you are doing to others? Have those concerns changed in recent months, and, if they have, how have they changed?

CHAPTER 2

Why should you study geography?

In the absence of geographical thinking, it is easy to overlook what is hidden and what is revealed when an issue is framed against the backdrop of a particular geographical space or scale. Studying geography, in other words, is a horizon-broadening experience.

— Alexander B. Murphy (2018)

THIS CHAPTER WILL EXPLAIN WHY you should study geography in the next stage of your education, whether as one of your GCSEs, at A level (or equivalent), as your university degree or even for postgraduate study. It will answer a number of key questions and provide guidance on where students and parents can find help in making the important choices that need to be made.

Why geography is worth studying

Geography, as we have already seen, is a subject that touches all aspects of human life. We're all living geography every day, and every day we make decisions that connect us with people in other parts of the country or in the wider world. This everyday (quotidian) geography is important, but to really understand the importance of these connections and be able to articulate why they are important, you need to 'study' geography in more detail. When I say you need to study it, I mean that you need to engage with the subject, spend time with it, do some reading and ask important questions in order to explore the new knowledge you are gaining through both formal and informal 'educational' experiences.

Don't expect to become an expert in all (or any) areas of geography immediately: remember that the subject has the whole world to explain and is a lifelong process of discovery and re-evaluation. The global hiatus in all of our lives in early 2020 caused by the Covid-19 pandemic led many people to think anew about certain aspects of how they live. One result of this was that it brought into sharp focus the many people we rely on to provide us with the things that are important to us, and who, in return, rely on us for their continued employment. This idea is called **interdependence**. Garment workers in India and Bangladesh were forced into strike action to protest about their low pay and unsafe working conditions. By working they would risk their lives; without their jobs, though, they would starve to death. Stories like this are also a reminder that geographers have to explore some dark places at times, such as Auschwitz and Chernobyl (geographers have a name for that phenomenon: 'dark tourism').

The enforced global 'anthropause' during part of 2020 was also a reminder of the importance of timing: we have become used to having the things we need readily available when we want them to be. The term 'anthropause' was actually coined by geography researchers, and an early example was the evacuation of the Urkainian city of Pripyat and its surroundings following the Chernobyl nuclear accident in 1986. Nature has now returned to the area, and parts of the city and the surrounding area have become a popular tourist attraction for those wanting something different to the usual.

The latest example of this phenomenon affected the majority of the world's population. Much of our daily lives are driven by convenience. Globalization has 'shrunk the world' and we are immediately and permanently connected. We can stay in touch virtually with our networks of friends and we are able to order clothing, books and entertainment for delivery the next day (or immediately) and to eat food grown in distant countries. The notion of 'social distancing' and the use of 'contact tracing' apps are profoundly geographical ideas that we have had to adjust to quickly.

In August 2020, Vicki Phillips, the chief education officer of *National Geographic,* wrote about the importance of a geographical education for the current times:

Understanding geography, the intersection of place and experience, allows us to explore thought-provoking solutions to this unprecedented moment in ways big and small.[18]

Logistics companies that supply both online and bricks-and-mortar retail companies work to strict timings, and many businesses run 'just-in-time' operations. Components and people need to be available at the right time or things can quickly go wrong: when they do, we can end up with disruption to production, growing food insecurity and, say, police officers guarding toilet rolls in Australia.

Timing is also significant to your reading of this book, of course. You're probably looking for some guidance on important decisions about studying geography or on what to do with your qualifications.

Some important choices to make

After spending a decade at school, you reach the first of several significant stages in your secondary education during year nine (K9 if you are reading this outside of the UK). You are presented with an important set of choices that need to be made. Having previously studied a wide range of subjects, you now need to make option choices for GCSEs (examinations taken at the age of 16) or their equivalents, narrowing down the subjects you are taught (with some being dropped: we presume you wouldn't have been so careless as to have dropped geography at this point).

Having made that difficult decision, a year into your GCSEs you are asked to make another choice: to narrow down your list of subjects still further to just three or four A level (or equivalent) choices.

A year into *those* courses, you are asked to make another choice: between undergraduate study at university (and between all the available courses if you take this path), an apprenticeship, a job or perhaps a gap year. You may also choose to take an Extended Project Qualification as well, which we will explain in more detail in Chapter 5.

Then, while at university, you can choose certain modules, and you will have decisions to make over dissertation and essay topics. Even if you don't go on to study geography at university, there is tremendous value added by studying geography beyond the age of 16.

If you plan to study geography at university, you'll have access to potential scholarship funds and loans, and ultimately to potentially greater earning power, but you'll also probably need to pay off substantial student debt. In the UK, repayments only commence once your earnings go above a certain threshold. It is worth finding out what the fees are for your own national context before making a decision about whether to go to university. The earnings power of geography graduates, as we'll see later, is better than most other subjects, although some of this is related to the school you attended: there are differences in the experiences of students attending state or private schools that need to be taken into account.[19]

Perhaps your plans for your future career have been guiding your subject choices from the beginning; you will, after all, need to secure employment. Ideally this will be rewarding and enjoyable and not too precarious. In the current economic climate, your career could involve many short-term jobs. A growing number of people have what is called a 'portfolio career': working for a number of different companies and organizations in a flexible way. During 2020, we saw the relative 'value' of jobs being questioned, with some lower-paid ones turning out to be essential for keeping the economy active and people healthy. There are also some immediate concerns about job security and the changing location of work. It's often helpful to include parents in these decisions, whether to gently guide you, share their personal experiences or help with your research (and dirty laundry).

An additional factor came to prominence in 2020/21 with the requirement for universities to teach remotely. Lecturers switched to Teams, Zoom, Blackboard and other group conferencing/webinar-style platforms. Some students actually benefited from these new ways of working. Those with social anxiety, for example, or those who find speaking in a seminar-style situation difficult sometimes preferred the relative anonymity of a webinar. Similarly, the ability to pause and rewind sections of online lectures was benficial to many, helping them to hone their skills of keeping up with the lecturer and being able to take notes alongside listening to new material. On the other hand, it is the social side of university that is key for many. Having the chance to debate in person, attend university social events and meet other like-minded people (and perhaps even a future partner) has been greatly missed.

The significance of leaving home, often for the first time, to live (semi-)independently is an important part of the whole university experience, as is learning about a new city and the geography of its local landscapes. A growing number of students live at home and study at a local university. Some people may have postponed their decision to start an undergraduate course as a result of the need for remote teaching (preferring face-to-face interaction), and others may have changed their minds on when to start a course. Disruption to overseas travel will have had an impact on those who had planned a 'gap year' volunteering or doing part-time work in other countries. Some will understandably swap the uncertainties currently involved in travel with the relative certainty of a three-year course of study.

As well as the actual academic subject of geography, there are the wider *geographical* implications of living in the first half of the twenty-first century, in an increasingly globalized and polarized world. The world faces significant challenges but also offers numerous exciting opportunities that geographers are well placed to take advantage of.

Are you at secondary school and thinking of studying geography?

At this point, you may be deciding whether geography at GCSE or A level (or equivalent) is for you. It might be helpful to hear some reasons why studying geography could be a powerful choice. Teachers will of course want you to study the subject they teach because they are passionate about it. They may already have showed you presentations in which they give reasons for continuing to study it. I hope they won't have resorted to saying negative things about other subjects: the value of geography should be obvious from their teaching over the years. One thing they may have mentioned is the fieldtrips you will have the opportunity to participate in. Fieldwork is a particularly important part of the subject: geography without fieldwork would be like science without experiments. Some of your long-term memories of school are likely to be made on geography field trips.

Avoid the traps of making choices based on what your current friends are doing or on whether you like the teacher you have at the moment. Having a friend sitting next to you in class will not improve your grade, and it won't be important in 15 years' time when you ask for a pay rise. Even if you allow your parents/guardians to advise you, make your own independent decisions: ones that you are happy with. The teacher you love may move on to a new job during your course, or they might end up teaching a different group to the one you are placed in. The friend who said they were 'definitely doing geography' could swap after two weeks of the course, or you might discover they picked something else anyway.

Advice is available from a wide range of places, including leaflets and weblinks from teachers. Some useful guidance on subject choices can be found on the Success at School website, on which there is also a geography-specific page.[21]

Ask your geography teacher which awarding body (previously called an exam board) they are following, as it might influence your choice. A list of UK awarding bodies is included at the end of the book, along with their contact details. They each have subject officers who look after the delivery of their subject. You may find it useful to visit the website of the relevant awarding body and read through their geography specification. This will outline what you are going to be learning about for the next few years. You can do this for other subjects too, of course.

Some geography GCSE courses use what is called an 'enquiry approach', which means you will have a chance to enquire into how things work a little more. Some specifications require more learning of case studies, and the nature of the assessment (examination) is slightly different too. Use the links to awarding bodies that are provided in Chapter 8 to take a look at samples of past examination papers as well: they will give you an idea of what to expect. Your teachers may also be able to provide past papers. Make sure you know what you are letting yourself in for when you make each subject choice. Sadly, you can't *just* take geography: you'll have to carry on with mathematics, English language and science courses whatever happens. When assessing your GCSE and A level choices you would do well to visit the website of the Royal Geographical Society, which has a specific advice area for those thinking about choosing geography at school or university.[22]

Although we return to careers in more depth in Chapters 3 and 4, it is worth saying here that schools and universities should provide students with independent careers advice that meets particular national standards. They should give you access to resources and employers, help with option choices and provide access to appropriate books (such as this one). Don't forget to use these resources to guide your choices. You may have also done some sort of psychometric test that suggested possible future careers when you provided your competencies, interests and motivations.

Other subject options may also include a geographical element, so you should carefully consider the combination of subjects that you choose. Geography pairs up well with most other subjects, as it has many characteristics that complement other fields.

Look for opportunities to get involved in additional geographical experiences. These can be helpful when writing personal statements later (see Chapter 5), but they'll also help you make the most of your course when it begins. Some examples are listed below.

▸ Join – or start up – a geography society or club at your school. This could involve students of all ages or be more focused on older students. It might involve reading groups, or movie nights with pizza, or it could be based around practical actions within your school: for example, it could concentrate on putting together a portfolio of evidence for the John Muir Award, which requires some practical action in your local environment.[23]

▸ Some schools may consider applying for the Eco Schools Award, which requires student participation and action by student-led committees as well as the development of an action plan for specific environmental improvements to be made to, or on, the school site. Get involved with the process or suggest that your school considers putting in an application for the award. Help gather evidence towards the first stage, which results in a Green Flag to fly proudly.[24]

▸ Find out if your school is applying for the Geographical Association's Quality Mark for Geography, and volunteer to help provide or coordinate the collection of student voices, or other evidence, that your teachers collate to pass to the moderators. The entry deadline is usually in June of each year, with moderation being done over the summer.

▸ You are likely to be offered a **work experience** placement sometime during your GCSE course, or possibly later. The timing varies from school to school, but placements often involve older students, who are able to access workplaces more easily. If there are options to choose placements, try to identify a workplace in which your geographical skills might be of more use, or where you may acquire them. Use parental contacts and the free careers guidance provided by your school or college to identify a suitable location and to check issues such as insurance and issues with access ahead of time. While on work experience, keep a diary of the experiences and skills that you have used or learned. This can be very useful when producing CVs and personal statements later on.

► Keep an eye out for talks by geographers and explorers at local theatres. People like Simon Reeve, Levison Wood, Ranulph Fiennes, Felicity Aston, Benedict Allen, Lucy Shepherd and many others are interesting and inspirational figures. If you see that they're talking nearby, you could mention this to your teachers: they may organize a group ticket.

► Find out about your local **Geographical Association/Royal Geographical Society** branches, as they tend to offer talks and lectures. They can be found in most regions, including in major cities. Your school may be involved in these: ask your teachers to find out for you.

► Read geographical books for your own interest. A list of suggestions is included in Chapter 8.

► Get involved with the **Duke of Edinburgh's Award** (www.dofe.org), which may be offered by your school or by another local provider such as a sixth-form college. The expedition element of the award is an obvious opportunity for developing your map-reading and navigation skills, but other elements of the award can also be turned to your geographical advantage: for example, the volunteering element could be carried out somewhere with a geographical connection. Students can start collecting evidence towards the award from year 9 onwards, and the skills developed during the silver or gold award levels in particular are recognized by employers as being useful in the workplace. Achieve gold and you could meet a member of the Royal family or one of their group of youth ambassadors as they hand over your badge. You will also find yourself in some unfamiliar and beautiful parts of the country doing practice expeditions and, later, assessed ones.

Although it would be preferable, it is not essential to have studied geography at GCSE level to achieve a high grade at A level, so it's worth noting that you may still be able to return to the subject if you realize later that you made the wrong decision not to opt for it when choosing your GCSEs.

Studying geography at A level

Geographers have a particular skillset that sets them apart from those who study other subjects, and this is intensified during A level courses. At this stage, students broaden their worldview and develop their ability to think geographically. The skills of geographers are very much in demand currently, and it's while studying for your A level that you can start to work more independently and begin to 'specialize' and dig deeper into areas of the subject you're more interested in. By this stage, you're likely to be carrying out more research. The complex ways in which people are exploiting the world's resources, and how global relationships are changing are two interesting areas of study. There are plenty of books and other resources that your teachers will expect you to explore once your A level studies start – these include journals and web-based material, which is added to on a daily basis. Geography never stops to take a breath.

Geographers develop a range of skills during their studies – both explicitly and less explicitly – that are also important at an academic level. A 2015 editorial in *The Guardian* described geography as 'the must-have A level', going on to say:

Geography is a subject for our times. It is inherently multidisciplinary in a world that increasingly values people who have the skills needed to work across the physical and social sciences.[25]

Data presentation tools and analysis for A level geographers
A level courses introduce students to a wide range of tools. Geographers have always made use of a range of tools to help them understand the world, and the number and variety of those tools is always growing. Google Earth has helped millions to see their house from above for the first time before flying off to explore other places in three dimenions. Here are a few other visualizations that might help you to see the power of this technology in helping people understand the world, and could also fire up your geographical imagination. You may be familiar with some of these, particularly if you are already studying geography at university.

earth.nullschool.net and Windy. You can see earth's atmospheric and ocean currents moving on the beautiful website https://earth. nullschool.net, created by Cameron Beccario. It will help you visualize and analyse a whole range of atmospheric data that is in 'near real time' (sometimes it is just a few hours old). By clicking on the word 'earth' you can see all the varied data sets that are accessible, and you can change the map projection. You can access visualizations of other weather data on a similar site called Windy: www.windy.com.

The Past Climate Explorer and the Climate & Energy Education Demonstrator. If you would rather look at historical weather data, make use of the Past Climate Explorer website at https://era5.lobelia. earth, which has a range of datasets. If, by contrast, you would prefer to look ahead, you can delve into climate change modelling by accessing the C3S Climate & Energy Education Demonstrator tool: https://c3s-edu.wemcouncil.org/. This was developed by the World Energy and Meteorology Council, and it was designed to allow students to handle data sets and explore projected changes.

Datashine. This tool, created at UCL, provides a portal at https://datashine.org.uk for exploring the huge volumes of data from the last national census (and updates that have happened since). It displays detailed information about all parts of the UK, and this is particularly helpful for exploring patterns of housing, say, or age demographics across cities such as London. Data of this kind are particularly useful at higher education level as well: they can bring to light patterns that geography can then explain. Change the contents of the dropdown boxes and zoom in to receive detailed maps of the UK that show important information. Similar tools, such as the European Statistical Atlas (https://ec.europa.eu/eurostat/statistical -atlas/gis/viewer), are available in other countries.

Parallel. This website offers a series of visualizations shared by a company based in Leeds: see https://parallel.co.uk/population. They include the excellent 'Population Estimates for England and

Wales' map, which provides the opportunity to instantly view a population pyramid plotted in seconds for any census output area. Something that used to take hours now takes seconds to achieve. Browsing cities like Leeds allows you to quickly identify the areas in which students at the city's universities mostly live. Similarly, if you head for the Devon or North Norfolk coasts you can locate the villages where retirees head for peace and quiet, with the average age rising into the 60s. You can also choose to explore air quality, fuel poverty and life expectancy.

Worldmapper. Worldmapper (https://worldmapper.org/) hosts a powerful series of hundreds of maps and cartograms that help to illustrate our unequal world by visualizing hundreds of data sets on health, education, industry and other geographical themes. The cartograms are all produced by Ben Hennig from the University of Iceland. Ben features later in this book as one of our 'guest geographers' explaining why geography matters to them. Once you have seen these cartograms your world view will change forever; they feature in many recent academic and school textbooks.

Flightradar24. Look up at the sky during the day and you will see contrails showing the routes taken by aircraft at altitude (although there were fewer of those during 2020 than at any time in recent history because of the grounding of aircraft during the global pandemic). The Flightradar24 site (www.flightradar24.com) provides details of every flight, including the type of aircraft and its flight path. The site can also be used to track the progress of flights carrying loved ones if you know the flight number; parents and guardians can use it to keep track of their children's travels. Browse the world to identify those areas where flights are fewer in number. You may be surprised just how busy the skies are: there can apparently be up to half a million people in the air at any given time.

MarineTraffic. This is a site similar to FlightRadar24 that uses the Automatic Identification System that is carried by all ships over a certain weight to track the movement of these vessels on the world's rivers, canals, seas and oceans: see www.marinetraffic. com/. You can search for ships owned by one of the larger freight companies, such as Maersk, and trace the movement of their vessels from Chinese ports, through the Suez Canal, bringing goods to the UK and the EU.

CDRC mapping. The CDRC is the Consumer Data Research Centre and it collects data that is then mapped for later analysis: see https://maps.cdrc.ac.uk/. The data includes information about the age of housing, the access people have to particular services, and other statistics on daily life, all drawn from consumer behaviour.

Climate Stripes. These were produced by Ed Hawkins of Reading University. They indicate changes in global temperature to show the warming trend. The first set to be produced visualized the global temperature change from 1850 onwards, and you can find stripes for different parts of the world on the associated website. Take a look at your own home area to see how things are warming: https://showyourstripes.info/.

Geography in the International Baccalaureate

In the terminology of the organization that coordinates and accredits the International Baccalaureate (IB), geography is classed as a 'higher level subject' (https://ibo.org). The IB is based in Geneva and runs four study programmes that all include a geography component. The two that are most relevant for preparation for university study are the Diploma Programme and the Career-Related Programme, both of which are designed for students aged 16–19.

The Diploma Programme includes a Theory of Knowledge course along with six other courses drawn from six different academic areas. Geography also forms part of the Individuals and Societies course. This is designed to help students develop a critical appreciation of

- human experiences and behaviour;
- the varieties of physical, economic and social environments that people inhabit; and
- the history of social and cultural institutions.

The geography course in the IB Diploma Programme integrates elements of both physical and human geography.

Geography as a subject is distinctive in its spatial dimension. It occupies a middle ground between social/human sciences and natural sciences, and it takes advantage of its position to examine relevant concepts and ideas from a wide variety of disciplines. This helps students develop life skills and gives them an appreciation of, and a respect for, alternative approaches, viewpoints and ideas.

Three topics are involved in the IB Diploma Programme:

- Power, places and networks
- Human development and diversity
- Global risks and resilience

These broad topics are made up of smaller elements that allow students to explore some important processes that shape the world.

The study of these topics is accompanied by fieldwork investigation: a written report that involves information collection and subsequent analysis, with evaluation of the collected data. The fieldwork report is a lengthy piece of work that requires some independent thinking and that therefore helps prepare students appropriately for further study.

The Career-Related Programme also includes a vocational element (that is not necessarily examined by the IB itself) combined with some Diploma Programme subjects. Students need to show they have acquired certain personal and professional skills. The vocational element normally includes a geographical flavour, often linking with the travel and tourism industry, for example. Some schools use outdoor instruction qualifications for this as well. There is a large amount of flexibility here to choose your own path. Some schools and colleges in the UK will offer the IB as an option for geography study.

Geography and gender

Geography actually has one of the best gender balances of any subject in terms of subject choices post-14. It's also worth saying that female students outperform male students in these examinations, and women are far more likely than men to earn more than the average salary as a geography graduate. For Russell Group universities, female students are also more likely to secure a first class degree than male students are.

There is some evidence that young women are slightly more likely than men to do an undergraduate degree that has an emphasis on human, rather than physical, geography. If you are interested in this aspect of your studies, you might want to read 'Gender inequality and women in geography', a report from the Royal Geographical Society that is available online.[26]

Young people and climate change

Climate change is the greatest challenge facing our planet. Geographers have been heavily involved in prolonged discussions on the topic and have given evidence at numerous meetings. They have been raising the alarm about climate change and other environmental issues for decades. When the Al Gore film *An Inconvenient Truth* was released, geographers provided specific commentary and guidance for teachers to refer to if they were using the film with students.

A survey conducted in September 2019 by YouGov on behalf of the Royal Geographical Society asked about the links between school subjects and the issue of climate change. The results from the survey's first question told us that geography was viewed by young people as the subject that was best able to explain climate change through the teaching they received. Geography GCSE was widely recognized as the subject that young people should study if they wanted to understand the effects of climate change. It also revealed that young adults (aged 18–24) are more likely to agree that studying geography can help you learn about climate change (82% of those surveyed agreed) than older generations are (72% of those aged over 55 agreed).

In the online poll, geography was ranked first among all subjects by both men and women, across all age ranges, social classes and regions, and regardless of whether the respondents had children or not. Geography was also by far the most popular subject listed, being selected by 74% of respondents – 30 percentage points higher than the second-placed subject (biology, which was selected by 44% of respondents).[27]

A further YouGov/Royal Geographical Society poll in 2020 led to the same results being seen for A level geography, with geology coming second as the subject that would best help explain climate change. Geography can help unravel the long-term effects of climate change, and it can perhaps provide workable solutions to mitigate the problem too.

Critical thinking

Geographers are critical thinkers by nature. There has been a growth in courses that explore these skills, but they need to be used in collaboration with a subject discipline. Geography provides those who study it with the antidote to fake news: geographers appreciate the need for serious evidence and critical analysis of the sources of information that they use. Images can be manipulated, statistics can be misused, and context is vital. Geographers develop the skills that are needed to spot errors and deliberate misrepresentations. The work of Hans Rosling's Gapminder Foundation is useful here in introducing the theme of 'factfulness' (a term coined by Ola Rosling). Factfulness is 'the stress-reducing habit of only carrying opinions for which you have strong supporting facts', and the book of the same name – which has been recommended by many and is influential in curriculum thinking – has become a catalyst for change.

Other opportunities to develop as a geographer while at school

The Royal Geographical Society runs a number of events each year that are worth considering, particularly if you live close to London or one of the society's regional venues. Students can join the Royal Geographical Society as a Young Geographer member, which is ideal for students, undergraduates and recent graduates. This allows access to the society's education resources and gives you the chance to attend a range of talks. Parents could consider a membership for their child as an investment in their geographical education.

The Geographical Association offers a Future Geographers programme at their annual conference. This is designed for A level students and their teachers, and it consists of a day of talks and workshops designed to develop subject knowledge. In recent years, it has included the chance to attend a special day at the conference, allowing participants to attend the Presidential Lecture and other sessions, to take part in guided fieldwork in the host city, and to carry out work in university laboratories in order to learn new skills connected with the subject specialism.

⁓✐⁓

Equipped with knowledge of the subject itself and its value, we may now be ready to ask: what is a geographer?

It's clear that geographers have a desire to understand the world around them, and to involve themselves in key debates about real-world issues. School can take you to a certain level, but university studies will develop your geographical thinking well beyond this. The next two chapters will focus on the world of university geography. They will take you through the process of applying for a geography degree and then describe how you can make the most of your undergraduate studies.

Some things to think about before reading the next chapter

▶ Start the process of finding out more about university geography courses by requesting a few university prospectuses. These can be downloaded in PDF format. Use the UCAS search engine (http://digital.ucas.com/search) to find the appropriate prospectus. You will probably be asked to register with each university and provide a few details and identify the year when you are thinking of starting the course before the prospectus can be downloaded. If you opt in for newsletters, you will receive regular updates during the period when you are making decisions, and these might also help you make your decisions.

▶ You could start investigating how student finance works and what additional financial support you might be eligible for from the Student Loans Company.

CHAPTER 3

Where can your geographical studies
take you?

We need among policy makers a much higher level of geographical literacy. We need people in positions of decision-making – both at national and local government level but also in business and in NGOs – to really understand the complexities of geographical systems. The decisions made in the next 15 to 20 years will be incredibly important for ... the whole of the UK.

— Nicholas Crane, RGS President 2015–18 (speaking in 2018)

IN THIS CHAPTER I WILL suggest some possible options for what your geography qualification will enable you to do, whether that is an A level (or equivalent) or a degree. The chapter will explore the great variety of options that are opened up for you by studying geography, and we will begin to identify some appropriate career paths.

I've used the phrase 'thinking geographically' several times already in this book, and it's time to consider it again now as we move onto the subject of which careers or jobs – and what other options – geography may lead you to. Unless you're particularly fortunate, you're going to have to secure a job after you've finished your studies in order to provide an income for living your life. Prospective students are now a little more career-minded than they used to be and tend to think more before investing their time in a university course (and accruing debt in the process). This is particularly true given recent periods of financial uncertainty. Tuition fees, which were already controversial at the time they were first introduced, were tripled for English universities in 2012. They certainly add up, as some of you reading this will know only too well.

Consider the following questions.

▶ What would you want to do with your geographical studies in an ideal world?

▶ Is there a job you've always wanted to do, and which studying geography could help you secure?

▶ What are your preferred future plans, in both the short term and the long term?

▶ Have you considered prolonging your academic study? Perhaps you might want to look for a geography postgraduate opportunity or complete a master's?

► Are there areas you have already studied that you think would make an interesting career?

Geographers are seen as particularly employable by many companies, and not just those that are an obvious geographical 'match'. You won't find many graduates with 'geographer' as their job title, and it's probably only teachers who could end up with the job title 'Head of Geography'. So while few employers advertise openings for 'geographers' per se – even when the job actually involves the application of geographical skills and technologies – people who have studied the subject to any level are, as I've said, particularly employable. Extensive professional possibilities are open to them, and they tend to have higher-than-average job security. They are often able to gain employment more quickly after finishing their studies than graduates from other subjects. Some subjects have an oversupply of graduates compared with the job opportunities available in that field, or directly linked to them. While a geography degree opens up a world of opportunities, a degree in dentistry, say, would probably lead to a life spent peering into other people's mouths.

Five years after graduation, geography graduates have an above-average likelihood of having secured employment, and their employment rates are better than those who graduated in many other disciplines, including politics, physics and history.[28] Geography graduates' average earnings are also higher than those in many other subjects, including technology and biosciences.

Although funding bodies and governments have shown a particular focus on STEM subjects (those in the science, technology, engineering and mathematical fields), and there is often a suggestion that graduates in these fields are the most employable, geography stands up strongly against these disciplines. Many geography degrees also align closely with some of those STEM subjects, as well as with the proposed SHAPE initiative that we'll discuss more later (see page 117). The Higher Education Careers Services Unit is clear that the knowledge, understanding, skills and approaches that one acquires during a geography degree are highly valued by employers. Over the last five years, geography graduates have reported very good rates of employment six months after graduation, and rates that are consistently higher than the average for all subjects.[29]

A May 2020 report by the British Academy and the London Economics Foundation provided another reminder of the value of studying geography.[30] Their report found that

▶ arts, humanities and social science graduates are now in greater demand than science graduates in eight of the UK's ten fastest-growing sectors of the economy;

▶ humanities graduates are as resilient to economic upheaval as other graduates, and they are just as likely to remain employed during downturns;

▶ humanities graduates are able to find work in the fastest-growing sectors of the UK economy, including in the fields of ICT and finance; and

▶ humanities graduates will also be essential to filling the workforce gaps of the future.

Here's what Hetan Shah, the chief executive of the British Academy, had to say at the time of the report's release:

We are increasingly living in a knowledge-based, creative and services economy. Arts, humanities and social science subjects are giving students the skills they need for the future marketplace. These graduates' flexible, adaptable skills make them extremely

resilient in the face of downturns and well-equipped to adapt to the technological changes in the job market. They go on to enjoy exciting and fulfilling careers in a variety of sectors including in those of enormous social value, for instance in the civil service, in teaching and social work.

Figure 1 Selected geographical careers followed by university graduates.

A report by the Institute for Fiscal Studies provided further important statistics about geographers and their earning power compared with other graduates. It seems they are more likely than those in many other subjects to be earning above-average salaries during their career. The Higher Education Statistics Agency publishes open data on student outcomes on its website (www.hesa.ac.uk).

Geographical courses may be found in the outcomes for physical sciences or perhaps social subjects, depending on the nature of the course. Graduate survey details can be found at the Graduate Outcomes website (www. graduateoutcomes.ac.uk).

Geographical opportunities are often found in places you might not have expected. A good example is seen if we look at the British Army's 42 Regiment (Geographic), a special regiment that forms part of the Corps of Engineers. They have been in existence since 1824, working as geographical specialists responsible for the production and supply of geographic information and products to a wide range of defence organizations. They create GIS-related products that provide key information for military commanders making mission-critical decisions. They also have a Reserve Squadron, called the 135 Geographic Squadron, that offers specialist training alongside the usual army training that all recruits receive. The first step in getting involved with them is a selection programme that includes a number of geographical skills tests:

- map reading and understanding,
- the effective use of geographic information systems,
- planning exercises,
- field terrain analysis exercises and
- presentation preparation and delivery skills.

For those of you with an interest in the services, this undoubtedly represents an interesting route in: one that would make the most of your geographical skills.

Where are all the geographers?

As mentioned earlier, there are few people who would list 'geographer' as their profession on an official form, but people with geography qualifications are employed in all sectors of the economy and use geographical approaches in their work every day. In 2019, the University of Bristol produced a report that outlined where more than 100 of their geography graduates had ended up several years after graduation: see Figure 1.

⤸

What follows is something of a pot pourri of examples, showing the different paths that a geography qualification can lead you down. It is by no means exhaustive but I hope it will provide you with some ideas for where your geography studies might take you.

Government

Since 2018, the UK Government has had its own Head of Geography. The incumbent is David Wood, and he has a responsibility to promote the value of geography in government departments alongside his existing job. He is supported by a number of Deputy Heads of Geography and other civil servants who have an interest in the subject.

Geography is now part of the Government Science and Engineering professions. These have been given an extra boost in recent years, and the government has taken more and more notice of geography in areas outside of education.

The Government Science and Engineering website provides the following analysis of the value of geography:

The work of government concerns people, communities and environments in specific places, often with strongly differentiated spatial characteristics. Analysing and responding to these patterns requires an understanding of how and why economic, social and environmental processes play out differently from place to place, at scales from local to global.

Geography and geographers offer government distinctive benefits. Geography is distinctive for its spatial analytical skills, for its ability to transcend scales, and for its integrating capabilities across multiple disciplines in the natural and social sciences.

Technological advances, e.g. the proliferation of locational data – such as that used and generated by mobile devices – has put geography at the heart of 'big data', adding to the already valuable toolkit of the discipline.[31]

The government's Treasury department uses what it calls its *Magenta Book.* This resource is used for formal assessments of government decisions, and for some time it has involved geography in its decision making. Certain decisions may be influenced by geography, whether cultural and political or physical in nature. Geography may have an impact on construction costs, for example, or the landscapes, geology and weather conditions of the route taken by a planned new piece of infrastructure might need to be considered. If you drive over the Pennines on the M62, the route chosen for the motorway avoids a higher, more-direct route because that would have meant the road suffered from a greater number of foggy days, increasing the chance of accidents. Keep an eye out for the farm (literally) in the middle of the M62 and see if you can find out why the motorway splits and goes either side of it.

Geospatial

During the Covid-19 pandemic it became apparent that mapping and tracking the outbreak was hugely important, as was tracking the movement of infected people. While geographical information systems (GIS) have been around for many years, the technology has become both more sophisticated and more easily accessible in recent years: not an easy double to pull off. It's also moved from being very expensive to being made free of charge for the online version.

The most widely used technology in this area is produced by esri, a company that was started by Jack and Laura Dangermond in 1969.

One of the skills that many undergraduate geography degrees will introduce students to, if they have not already been exposed to it at school, is the use of GIS. The most common tool for people to use is esri's ArcGIS. Most schools in the EU, including all schools in the UK, are given free access to the ArcGIS online software for all students and staff. You can even sign up for a free personal account and start making maps immediately. An important feature of GIS software is that it is possible to ask questions (queries), with the answers being displayed on the screen, often in the form of a map.

esri say of geography that it is:

the science of our world, providing a way to organise and integrate our knowledge, helping us to interconnect every discipline. GIS provides the technology for abstracting geography and a language for organising and applying geographic knowledge.[32]

A major area within which many geographers now work is called the geospatial industry. More than 10% of British industries are reliant to a great extent on map data, and location data is vital to *all* industries in some way, whether it relates to the location of customers or raw materials, say, or to calculating the most cost-effective way to deliver products by selecting the location for a new distribution warehouse. Parcel tracking, component ordering and the transport industry all use locational data to reduce costs and streamline operations. The geospatial industry is growing in size and economic value all the time. Geodemographics, which helps to explain the behaviour of large numbers of people, is another area of growth. This is a sector of the economy that is worth many billions, and in which there are good job prospects. Geographers are most welcome here because of their geoliteracy and their spatial thinking skills.

One fan of this technology is the musician Will.i.am, who has previously talked about how his teachers helped him escape the area of Los Angeles in which he was born and steered his life in a different direction. The story of his discovery of the power of GIS, and of how he now helps other young people discover the power of maps, can be found on the Teach with GIS website.[33]

Spatial epidemiology has become a topic we all now have an interest in, of course, and the contact-tracing process essentially uses the same techniques that John Snow used in the 1850s to track down the particular pump in the Soho area of London that caused a cholera outbreak (proving, in the process, that cholera was a water-borne disease). The data dashboard created by Johns Hopkins University to track Covid-19 cases has been viewed hundreds of millions of times.

Figure 2 Covid-19 Dashboard by the Center for Systems Science and Engineering (CSSE) at Johns Hopkins University (JHU) (https://coronavirus.jhu.edu/map.html).

In 2019, the Royal Society published a report called 'Dynamics of data science skills'. It suggested that in order for the UK to meet the needs of employers and maintain its position as a leading data research nation, it needed to take action in four areas. Further study in geography will prepare you to meet all four.

▶ **Develop foundational skills:** ensuring our education system provides all young people with data science knowledge and skills. This will require curriculum change within ten years.

▶ **Advance professional skills and nurture talent:** offering nimble and responsive training opportunities and developing training based on collaborations between the academic, public and private sectors.

▶ **Enable the movement and sharing of data science talent:** addressing barriers to mobility between industry, academia and the public sector.

▶ **Widen access to data in a well-governed way:** opening up data securely and providing access to computing power.

The Royal Society also suggested that to help with these four areas, GIS should be incorporated into the school curriculum – alongside other data analysis skills – more than it already is. Geography is the best-placed subject to offer these much-needed skills, and if you haven't already been introduced to GIS, a university geography course is guaranteed to remedy this. The main report is accompanied by career profiles that may provide further guidance.[34]

Google has its own geographer: a mapping and technology expert called Ed Parsons. He is clear that geography is a vital science for everyone to be able to access and that we all need to understand the power of geography. He has referred to Google Earth as 'a browser for the planet' and has spoken extensively about the importance of geospatial data. His job title at Google is Geospatial Technologist, indicating the importance of location to the company. During the 2020 lockdown we saw Google using smartphone data as part of 'track and trace' systems for apps, allowing outbreaks to be rapidly contained using information about the proximity to other people of those known to be infected.[35] Geographers appreciate the ethical issues surrounding the collection of such 'big data', particularly when it could be used to identify individuals and their movements.

Joe Smith, the director of the RGS, spoke to *The Guardian* in 2019 about the rapid growth of the geospatial sector, explaining that

> *the cabinet office's geospatial commission estimates that UK plc can gain £11bn through the better use of the geospatial data held by organisations including the Ordnance Survey, Land Registry and the Hydrographic Office. Yet this can only be achieved if we have enough graduates trained to understand how such data can be used to benefit our communities, environments, businesses, and policy decisions – training which is provided through the study of geography.[36]*

This is certainly an area to consider, then. It is always worth expanding your geographical curiosity.

Those who can, teach
One obvious job that geography study can lead on to is the chance to teach the next generation of geographers. A major advantage of teaching is that every day offers you the chance to explore different aspects of the subject. If you really love geography, there are few jobs that allow you to engage with the full range of topics rather than specializing in one area. Good teachers also continue to engage in the discipline, and the curriculum they teach is in a continual process of 'becoming'. In recent years, the Department for Education has provided bursaries to those who train in particular subjects, including geography. For 2020/21, the bursary was £15,000:

a higher amount than is offered to those training in other humanities subjects such as history and religious studies.

The RGS-IBG has created another scholarship scheme that offers additional support to teachers who are accepted onto it, following an interview process. The range of options for training as a teacher has broadened significantly in recent years, and there are now many more way into the classroom than used to be case. There has been a dramatic upsurge in interest in teaching as a career in 2020 – and the holidays aren't bad either.

Further geographical research and postgraduate opportunities
Geography is a subject that provides numerous opportunities for research, dealing as it does with a wide range of physical environments and human and cultural contexts. Options for further research in geography include studying for an MA/MSc/Med (a master's degree) or undertaking a PhD/DPhil or a similar research post (a doctorate).

Just as the range of undergraduate courses is wide and varied, there are similarly interesting options for master's-level research and for funded PhDs. Further study delays the arrival of employment but may eventually open up more job opportunities. In June 2020, it was reported that geography postgraduates are also more likely to be in employment than those from other subjects.

I've been fortunate to work with a number of geography researchers over the years and they always bring an interesting perspective to the work they complete. Some of this research finds its way into the classroom when teachers engage with academics.

Postgraduate opportunities in geography are as varied as they are in other areas of academic study, with the variety of topics that are in geography's orbit perhaps even greater than is the case for undergraduate study. A master's qualification will open up further funded opportunities in universities that are keen to attract research grants and need the brightest and best geographers. The Research Evaluation Framework, which assesses the 'value' of research, looks for evidence of impact, which geography courses often have numerous opportunities to demonstrate.

Geopolitics and diplomacy
In 2011, Madeleine Albright, former US Secretary of State, said:

> *Geography played a leading role in nearly every policy decision I was involved in as Secretary of State. Young Americans with an understanding of peoples, places and cultures have a clear advantage in today's rapidly changing global economy.*

Geographers understand the geopolitical situation in the many parts of the world that are contested and that need some diplomatic support to maintain the status quo. Critical and cautious thinking is required to avoid flashpoints leading to global or regional conflict.

Planning, and the design of places
Geographers have always been involved in the planning of places for people to live in, whether that be buildings or entire cities. Place itself is an important aspect of modern geography, and the relationship between buildings, infrastructure and the needs of people is well understood by planners. Good design can help reduce crime rates in a housing estate, for example. Paths that are lined with vegetation and/or without decent lighting are less safe for residents, and they may restrict some people's movement after dark.

After World War II, when much of the centre of London had been destroyed during the Blitz, geographers worked on the plan for how the new city should rise out of the rubble. Today's city of London is built on that particular period of history, which was in turn built on previous layers.

Transport planners work to develop the infrastructure that keeps our cities moving. The simple geographical models of Bradshaw, Burgess or Clark and Fisher that you may remember from your school days are transformed into far more complex modelling to work out solutions to urban problems. The Mayor of Paris Anne Hidalgo has promoted the idea of the *'ville du quart d'heure'* ('the quarter-hour city'), with the aim of offering Parisians everything they need either on or near their doorstep. This could trigger an ecological transformation of the capital, turning it into a collection of neighbourhoods within which people could cycle or walk. This would reduce pollution and stress, and help create socially and economically mixed districts, which would improve the overall quality of life for both residents and visitors. You could, therefore, find yourself working to shape and improve the lives of millions of urban dwellers in the future. The lockdown in Paris has allowed many roads to be transformed into cycle lanes to help jump-start these ideas. The Australian city of Melbourne is developing '20-minute neighbourhoods' in a similar project, and Amsterdam is pairing up geographers with economists, transport planners and other experts to work on something called 'the doughnut' to help with sustainability measures.

Geographers can help with projects like this because, as Alexander Murphy says:

> *an understanding and appreciation of geography should not be seen as a luxury; it should be viewed as vital to the effort to create a more liveable, just, sustainable and peaceful planet.*[37]

Insurance analysts

Insurance companies use geographical data in order to explore the nature of risk in particular places and to identify particularly risky locations for industrial activity that is vulnerable to disruption from hazards. Postcode-linked data can tell you a lot about an area, and the profiling of neighbourhoods is also done for the purpose of identifying risks.

One of the largest insurance companies in the world is Munich Re. They produce a number of reports each year that establish the nature of the risk facing different sectors and locations, and they use information about the changing nature of the world to set premiums for those who want to insure against particular risks. Geographers can help identify locations that are at particular risk from hazards: tectonically active areas, for example, or those that are prone to tropical cyclones.

The Discover Risk website (www.discoverrisk.co.uk) is a great resource for jobs in the insurance sector.

Oil exploration

Oil companies have always used geologists to help locate rocks that are potentially oil-bearing, and this is linked to the interpretation of mapping and subterranean surveying. Geographers may have a geological element to their degree or choose to complete a joint degree. Construction projects will also need geographers/ geologists to conduct ground and soil surveys on stability for road embankments, bridges and the construction of dams. These are highly specialized jobs requiring advanced subject knowledge.

Geography and ... something else

Geography has overlaps with many other subjects, and this means that there is likely to be a geography course that will suit both your interests and your skills. An interest in economics, chemistry or even sociology would still allow you to find a career with a geographical focus to it and make the most of your degree studies. Table 1 (overleaf) shows the ways in which various kinds of geographical content and study overlap with an array of other disciplines. Each of the areas of study in the 'Regional geography' column remain geographical in nature because they contribute to geography's core concepts.

Table 1 Geography overlaps with other subjects.

Regional geography	Systemic geography
Behavioural geography	Psychology
Population geography	Demography
Cultural geography	Anthropology and sociology
Medical geography	Health sciences
Economic geography	Economics
Historical geography	History
Urban geography	Planning and urban studies
Climatology	Meteorology
Geomorphology	Geology
Marine geography	Marine science
Biogeography	Biology
Political geography	Political science

Table based on an original figure in *Geography: Realms, Regions and Concepts* by H. De Blij and D. Muller (Figure 6-15). Wiley (2020).

Flooding and coastal engineering

Flooding is the most common 'natural' disaster and the one that's most likely to affect you personally in the future. A growing effort is being put into engineering that reduces the risk of flooding. The terms 'mitigation' and 'adaptation' are important here, and geographers are well suited to understanding the best places to locate particular defences. Coastal management schemes have a limited lifespan, which means that this field offers a career with longevity. The Environment Agency is the government body responsible for overseeing this work; it employs more than 10,000 people in a wide range of roles.

The Civil Service

The Civil Service delivers public services and supports the government of the day in the development and implementation of its policies. The work that civil servants do touches all aspects of life in the UK, from education and the environment to transport and defence. Civil servants help ensure the effective running of government and the provision of the best possible services to the public. This important work requires geographers, of course. They have an analysis function, including in the government's geography function, so it's possible to join the Civil Service as a geographer: there are roles as analysts, advisers, geospatial information specialists or multidisciplinary geographers, among others.[38]

Cartography

Geographers use a wide range of tools. This doesn't necessarily mean physical tools, although a compass and a paper OS map are the best tools for navigation, assuming you have some knowledge of how to use them. Mountain rescue teams always advise against a reliance on a smartphone and apps. Geographers also make use of a wide range of data, with a key data source being what is called Open Data. Open Data can be freely used, reused and redistributed by anyone, and are not 'owned' by a company who might charge for their use.

The speed at which the world changes is remarkable. The OS tries to keep pace with changes to the geography of England, Scotland and Wales through a number of avenues: they employ state-of-the-art GIS tools to plot the location of millions of 'assets'; they have a team of surveyors who criss-cross the region they work in; and they use a small fleet of planes that fly during the times of the year when there is less cloud and longer days in which to photograph the country. The result is that around 10,000 changes are made to its 'master map' every single day. You can follow the OS planes on the FlightRadar24 website during the summer months. They have the ID letters G-TASK and G-FIFA, and their work is described in a blog with the title 'Chocks away – tales from Ordnance Survey's flying unit' on the OS website.[39]

One major bugbear is when people say that geography is 'just colouring in'. There is certainly an element of this in some aspects of geographical analysis: maps tend to be coloured in, for example, with choropleth shading being a useful mapping technique for iden-tifying patterns in data. It's also helpful to use standard colours to distinguish which bits of a map are land and which are sea; similarly, pie charts use colours to distinguish between the different segments. The reality is that one of the most important aspects of the subject is the ability to **think geographically**, and colour can be helpful with the process of classifying what is in front of you, so that you can then concentrate on applying your knowledge.

In June 2019, geographers at the Bartlett Centre for Advanced Spatial Analysis (a research centre within UCL) launched the Colouring London website (https://colouringlondon.org), which invited people to do just that: apply colour to every building in London. Anybody who took part would have quickly discovered that there was, of course, a methodology behind the colouring process: its intention was to use imagery and data about each building to connect those engaging with the website with the 'poverty maps' of Charles Booth, the ongoing changes in the city of London, and the building materials used to construct the different areas of the city. Open data was provided about the use of each building, and about its age and context. It's a reminder that cities are dynamic and generally develop organically over centuries, with different time periods seeing their own additions being made. It's also a reminder that colouring in can reveal hidden depths.

Consultancy

Geographers are important in many consultancy roles as well as permanent posts. They may be called in to provide advice for specific projects, as they develop specific expertise in areas that are broadly useful for many large-scale developments. The statistics we mentioned earlier show that geography is particularly popular right now. Perhaps it is having its moment, and the pandemic has reminded everyone that geography really matters and will continue to do so.

How to obtain further experience of geographical jobs

If you're planning to dip your toe into the waters rather than plunge straight into full-time work, a number of options for gaining experience are available. The following ideas might help you decide on a future career direction.

Internships

An internship provides an opportunity to gain some insight into a company by working there for a limited time. It might come with the opportunity to earn money, either straight away or later, as the role develops; or it might simply be a way of 'getting your foot in the door' in the hope that future opportunities might be offered to someone who is already known to the company. If you take up an internship, look out for the possibility of volunteering for other related roles, as this can often lead to the development of new skills.

Aid agencies

Aid agencies require volunteers and paid staff with experience in geospatial thinking.[40] They mostly work in particularly risky areas of the world. The risk may come from a disaster that's considered to be 'natural', such as an earthquake, a tsunami or a tropical cyclone, or it may occur as a result of human-induced famines.

Shelterbox (www.shelterbox.org) is a UK-based charity that works globally. I spoke to Céline Chhea (the company's logistics assistant) a few years ago, and she told me how she used her geography qualifications to help plan the logistics of aid after a disaster:

Google Earth is sometimes used to get a 'feeling' for the terrain which surrounds an area. We also use Null School's Earth visualization tool to explore the winds that might be affecting the area during the time we are there. Winds may affect access to particular airports or generate swell for maritime vessels. High and low tides for docking at ports and working out which vessels of particular size and draft can be used from those which may be available to us.

We are juggling the weight and size of particular items to ensure that they are carried on the most appropriate vehicle. Our plans change as circumstances change – we have to be flexible.

Céline also told me about working in Malawi following a major flood. Floods had previously affected the same area, along with earthquakes, and Malawi is therefore considered a disaster hot spot. She told me how geography helped with the distribution of aid:

Neighbouring countries are unstable, so there could be an issue with getting aid in by road. Borders are in conflict areas – and within the country there are areas which are safer than others. There is a real risk of debris compromising water quality – which makes the provision of potable water a priority. Electricity supply availability and disruption is another factor, as flooding can disrupt this. Weather forecasts for the next few weeks are required if possible, and certainly short-term forecasts are essential to ensure that we could assess which areas might get worse. Knowing how rivers flow is also important here, so physical geography knowledge is relevant. The ground can change over time, as river levels are swollen by further rainfall. Topographical mapping and digital elevation models (DEMs) are used to get accurate terrain data. We need to know: where is the higher ground to move to if required as flooding worsens, or to operate from to minimize disruption to operations?

Water bodies were also important in countries that are landlocked or otherwise limited in accessibility, as boats are sometimes used in these situations.[41]

Why don't you read the quote again and look out for how often geographical themes are mentioned in what Céline says? You can find out more about Céline's job and read the full interview online.[42]

Volunteering in geography

You might not have considered volunteering, but it can provide you with further skills and insights that will help you choose your future career path. It's a great way to build up your skills and experience, and it can sometimes pave the way for a paid position, either within the same organization or by giving your CV a significant boost. Some voluntary roles or internships can lead to qualifications such as National Vocational Qualifications or certification from local colleges. They widen your network of contacts and experiences and enhance employability. Surveys suggest many employers are more likely to recruit candidates with volunteering experience over those without any. Further information can be found on the Do-It website (https://do-it.org/).

There are plenty of UK opportunities that have a geographical dimension, including environmental and conservation work with local wildlife trusts, the Royal Society for the Protection of Birds (who have nationwide opportunities), the CPRE ('the countryside charity'), the National Trust and the John Muir Trust. Another alternative is to try to find a placement in climate change or development education. Geographers often have a desire to improve the environment because they understand the many threats that it faces, on both a local scale and a global one. Again, the Do-It website is a useful resource, as it has thousands of voluntary opportunities that you can search according to location and interest – art and culture, environment, heritage, international aid, museums, and sports and outdoor activities should all appeal to geographers. Time spent volunteering outside can be good for your mental and physical health, and it will allow you to pick up some practical and 'soft' skills, such as teamwork.

Businessman James Caan, one of the former 'dragons' from the BBC's *Dragons' Den* programme, has said:

> *When I see two graduates whose CVs read the same, I look for activities like volunteering which gives the CV a unique selling point. As an employer, listening to a graduate talk about volunteering means you can start to see their personality and [it] helps you to visualise them in your company.*

Another helpful UK-specific website is that of Vinspired (https://vinspired.com), a government-supported national charity for those aged 16–25 that encourages people to participate in voluntary work placements. The organization works with others to offer a wide range of placements, some of which include accredited training and cover living expenses. As the website says, this is a great way to 'Make a difference, Make your mark'.

A particular area of interest for geographers is in conservation volunteering: see the website of The Conservation Volunteers (www.tcv.org.uk). You might start with 'day volunteering', working your way up to longer-term placements. The organization has been active across the country for more than 60 years. Global Vision International (www.gvi.co.uk) is another great resource, offering a range of options.

It is also worth finding out about volunteering opportunities for while you're still studying, as many universities have societies that encourage participation in such schemes. This could be a way into an exciting future career in a geographical field – or in an actual field.

Studying geography overseas

Studying geography provides plenty of opportunities to 'go to new places', both in terms of physical travel and in terms of education, employment and career development. Overseas travel is also part of many geography degrees, with fieldtrips outside your home country being undertaken on most of them. The actual locations you visit will depend on the research interests of your university tutors and lecturers, but nowhere on earth, or even beneath it, is off limits. Look out, too, for opportunities that arise through interest groups and societies, and explore the possibility of overseas study with reduced tuition fees. UCAS provides further details and a list of overseas providers (www.ucas.com/undergraduate/what-and-where-study/ studying-overseas).

Studying geography at university is, for many, a chance to develop both social skills and very practical ones while living away from home for the first time. Further travel may also be made possible by the nature of the course itself. It's worth remembering that many universities in the EU also offer teaching in English, often with fees that are lower than those for some UK universities.

The ERASMUS scheme and other similar initiatives are provided by the EU and by other organizations. They are designed to allow students to move around internationally, to access training and work with others, and to develop themselves professionally. Brexit may affect the opportunities for UK students to access these schemes, but the position of ERASMUS itself remains uncertain. I certainly hope that it will remain something that you'll be able to take advantage of. It is easy to find stories of people whose lives were 'made' by involvement in the scheme.[43]

Apprenticeships

Recent economic challenges have led to additional funding being made available for apprenticeships, and the number of them is therefore growing. They have become more accepted as an alternative to undergraduate study and they may be more appropriate for many because of their vocational-training element, and because they provide a route into full-time work. There is a dedicated area of the RGS-IBG website that provides guidance on apprenticeships, although readers are reminded that there is no dedicated 'geography' apprenticeship, as they are organized according to occupational groups. Apprenticeships are offered at a number of levels: from 'Intermediate' (level 2, equivalent to GCSE), to 'Advanced' (level 3, equivalent to A level), right up to 'Higher' (levels 4–7, which offer degree-equivalent qualifications).[44]

At the time of writing, we're in a period of high employment (although the definition of employment has been changed, so that as little as one hour per week of paid work means you are classed as in employment!), but many UK workers were furloughed because of Covid-19, and job losses became increasingly inevitable. Some changes to the numbers of apprenticeships and the available funding for training have come about because of the pandemic, so it's a good idea to be flexible with your plans and choices. You can be confident, though, that geography can give you an edge in the job market.

There are a range of apprenticeships linked to geography. Some examples are found in

- agriculture, horticulture and animal care, e.g. as a ranger or a woodland manager;
- leisure, travel and tourism, and logistics, e.g. as a groundsperson or a travel adviser; and
- construction, planning and the built environment, e.g. as a surveying technician.

The government will need to invest in the economy and in workforce training in the coming years, so there are likely to be additional options available to you. A dedicated website for apprenticeships called the Apprenticeship Guide includes specific subject areas and is a valuable resource.[45]

Apprenticeship opportunities come out throughout the year, so keep an eye on areas that you are interested in. The government's 'Get In and Go Far' campaign has a national website that advertises vacancies (www.apprenticeships.gov.uk), and it is also worth checking for local apprenticeship providers, as students may be able to access resources and advice that way.

Virtual work experience

A new area – and one that's likely to grow and become more significant – is that of opportunities for virtual work experience. Also referred to as online work experience, remote work experience or e-experience, virtual work experience gives students and graduates the opportunity to complete a full **internship** from home, needing only their laptop and an internet connection. Virtual internships share many similarities with traditional, in-person internships, but the fact that they're online allows participants to work with organizations around the country, or even the whole globe. They're particularly relevant in these unprecedented times, as they enable students and graduates to gain invaluable work experience and boost their CV in spite of lockdown restrictions. More details can be found on the Prospects website.[46]

Joint degrees with geography include a number of openings into particular careers

Geography can be combined with a number of other subjects. Some of these combinations have grown in popularity in recent years, such as the links with ecology. Ecology is a field of study that allows you to engage with natural environments while carrying out your research. There has been a recent increase in the level of engagement with the 'new' nature writing of authors such as Robert MacFarlane, Melissa Harrison, Tim Dee and Rob Cowan. There is a recognition that we need to protect the natural world, and just as construction projects need geologists, they also need to include environmental surveys in order to reduce their impact on ecosystems and groundwater.

Plymouth University is an example of an institution offering degrees in geography with international relations. It has designed modules that address many of the Sustainable Development Goals of the United Nations (www.un.org/sustainabledevelopment). These goals – which replaced the Millennium Development Goals in 2015 – were developed as the 'blueprint to achieve a better and more sustainable future for all'. They are ambitious, aiming to kickstart a concerted effort to reduce various global problems. All countries have been challenged to make progress over the next 15 years towards achieving the goals, which have been described as

a call for action by all countries – poor, rich and middle-income – to promote prosperity while protecting the planet. They recognize that ending poverty must go hand-in-hand with strategies that build economic growth and address a range of social needs including education, health, social protection, and job opportunities, while tackling climate change and environmental protection.

As a geographer, it's worth exploring these goals and becoming familiar with them. They underpin a lot of the work that your teachers complete with you in school, and they are relevant to a great deal of undergraduate study.

The Plymouth University course also offers an optional year in work placement, developing further skills.

If you've decided that you're going to have a gap year ...

How exciting! Many people want a break before or after their university studies, and parents and guardians reading this book may also remember their own gap year travels.

Gap year travel provides an opportunity to broaden your world-view, but you should also return with a fresh appreciation of your own home area. Rudyard Kipling wrote, 'What do they know of England, who only England know?' You will return with a better understanding of the connections that your home area has with these distant places, your geographical imagination will have been broadened, and you may also have developed some useful language skills and possibly even made contacts that will help you enter the world of work.

Think carefully about where you are going to travel, how you will travel, and who you will travel with. Technology has opened up the world to everyone and has allowed travellers to stay in much closer contact with family back home, even if, for now at least, there are fewer options for travel, particularly across national borders.

Spending time in other countries, particularly outside Europe, has the potential to change your life plans completely depending on chance encounters on your travels. Keep an eye on travel guidance from the Foreign, Commonwealth & Development Office (formerly the Foreign & Commonwealth Office, or FCO), but do think about exploring the opportunities for reduced-price travel for those who are still young. Start your preparations as soon as possible and do your research carefully.

Have a look through some of the media resources listed in Chapter 8. Why not produce your own geographical top ten of places that you have experienced and then share it with your teachers to let them know where you've been? I've received postcards from former students from locations all over the world, often being told that they were inspired to visit a certain place after hearing about it in a lesson.

Hunt out your passport (burgundy or otherwise) and make sure you have decent travel insurance. Because of the Covid-19 pandemic, travel is not going to be as straightforward as it used to be for some time, so carefully consider how and where you intend to travel, and always have a 'Plan B'. Remember that you may need visas to visit certain countries, depending on your own nationality, and you also often need to be able to prove that you've been innoculated against certain diseases.

Plan your routes carefully and avoid missing the geographical highlights of the cities and countries you plan to visit by always being ready to stray from the normal tourist trail. Learn a few key phrases in the languages of the countries you are visiting. It may help you to get out of trouble, or just to locate the nearest toilets.

You may want to seek out ethical travel and volunteering groups abroad, to reinforce the sustainability elements of geographical study. There are several websites offering the chance to find out more before committing yourself: the Working Abroad website has some good advice, for example.[47]

It is worth being aware of the fact that there have been some issues with the way that this kind of activity can be perceived – the phrase 'voluntourism' is sometimes used pejoratively. Investigate opportunities such as those provided by Pod Volunteer (www.pod-volunteer.org), but do your research if you really want to make an impact on communities overseas.

Some things to think about before reading the next chapter

► Consider your own circumstances and what you would be comfortable doing as a future career. If a university place is part of your plan, for example, can you easily picture yourself living away from home? What size(s) of city would you be comfortable living in? What are your plans in the longer term?

► Check out the Prospects website, which suggests a range of things that you can do with a geography degree.[48] Search the #ChooseGeography hashtag on Twitter to find tweets about the value of choosing geography.

CHAPTER 4

How have people used their geography qualifications?

It is geography that applies common sense to the statistical hysteria of the climatologists. It is geography that brings global warming into context and applies the test of feasibility to whatever political priorities are deemed necessary. It is geography that explains why each of us is located where we are, in neighbourhood, nation, continent and planet, and how fragile might be that location. Without geography's instruction, we are in every sense lost – random robots who can only read and count.

— Sir Simon Jenkins (2017)[49]

THIS CHAPTER EXPLORES INDIVIDUAL STORIES from some people I've spoken to personally, and from others who have taken their geography qualification in an interesting direction. We've seen already why geography is important, and discussed its value to a range of future careers, and this section fleshes out those ideas with the personal experiences and observations of a number of people whose career uses geography in some way. Along the way we'll hear why people believe geography is vital to explaining our lives.

I've already explored why university geographers tend to stay the course, and I've explained why geographers are more likely than average to secure a job soon after their graduation. Employers increasingly value globally minded staff: those who understand intercultural differences and see the opportunities they present for innovation and change. Broad-minded people who can see beyond the obvious are highly prized.

Here are stories from a range of people who have used their geography qualifications in different and varied ways.

Kate Edwards

Dr Kate Edwards works in the field of game design: a passion that started with her early interest in video games and Middle Earth. She has since become a consultant for Microsoft, helping them ensure their products are culturally appropriate for the markets in which they are sold. In her Digital Worlds lecture at the Royal Geographical Society in 2019, Edwards outlined the power of these games to reconstruct the world in a new form. She explained how people have been creating new worlds for as long as they have been creating anything. The layers that need to be applied to make a world 'follow the rules' are similar to the way that GIS works: extra information on top of a base map.

Geographers are particularly useful people to involve in the work of games design companies, as we spend a lot of time trying to understand the real world. Kate suggests that when we see a map, we trust what it is: it establishes the world of a book or game as being 'reality'. Some games need a realistic climate, topography (and geomorphology), biology and even political system to draw players in. The worlds in games such as *Assassin's Creed* and *Call of Duty* are realized in tremendous detail. Creating them involves many decisions about what to include or exclude in order to produce a consistent world that those playing the game can believe in, and ensuring that no offence is caused to cultures and religions that are portrayed in a game is very important. Tens of millions of pounds can be spent on game development.

Edwards says of her job:

The role that geography plays in global business, and especially in the technology sector, is so implicit at times that we have to be resourceful and innovative in how, where, and in what role we apply our valued-added knowledge. Very few companies will ever have openings for a 'geographer', so with guidance from mentors in the tech and video game fields ... I was able to discern how my skills and perspective could benefit companies.[50]

Iain Stewart

Iain Stewart is a professor of geoscience communication at the University of Plymouth. He has presented a number of TV series including *The Power of the Planet* and *How Earth Made Us*. The following quote from 2014 makes clear his thoughts on geography needing to make the most of its interdisciplinarity.

This notion that the best thing to do is to separate our study into the human bit and the physical bit is not fit for the 21st century.

He has travelled widely and speaks on the value of communicating geography to people, so they recognize the importance of understanding how tectonic forces make settlements vulnerable, for example. He reminds us that while many people might think that earthquakes kill people, it is buildings that kill people. The way in which people have allowed them to be built, using particular methods and materials in earthquake-prone areas, is the problem.

Joe Smith

Professor Joe Smith is the current director of the Royal Geographical Society with IBG. The society has been in existence since 1830, with the remit of promoting geographical science, notably through exploration but more recently with an increased focus on education. One might assume that Joe did a geography degree, but he didn't in fact start out with that as his specialism, as he told me:

After studying a Social and Political Sciences degree at Cambridge I knew I wanted to parallel my engagement with environmental politics along an academic and policy track. The people in my department were great – world leading indeed – but mostly couldn't spell the word 'environment', and that led me to the door of the geography department in Cambridge, and study for a PhD with Professor Sue Owens.

That word 'policy' is important here. Geographers are often involved in policy roles, partly because of the way that geographical ideas are so tied up with most important decisions. This opens up a range of job opportunities for geographers, as we have already seen. Joe continues:

It was immediately obvious that geography departments were the only place that knew how to bring together the natural and social sciences and humanities to try to make sense of complex problems. They didn't have to go to an institutional drawing board to 'make up' interdisciplinary working: it was in the DNA of the discipline. I loved the extraordinary scope of intellectual ambition all around me, which was inspiring.

We have already seen that geographers are good systems thinkers and analysts. We try to build up 'whole pictures' when we approach a problem. As Joe told me:

In addition to providing a good intellectual basis for strategic and system-wide thinking, geography is simply relentlessly interesting, and that keeps you going in any career.

The career profiles on the RGS-IBG website provide a really diverse menu of opportunities for people that have studied geography and would like to carry their interest in the subject through to their career.[51]

Joe and I also discussed Chartered Geographer status, which has been in existence since 2002. A team of moderators assesses applications that come from people who have been using geography in their careers for a period of time. This professional qualification joins the likes of Chartered Surveyor, Chartered Accountant and Chartered Engineer, among others. Those holding the status have to reapply annually with fresh evidence of their activities.

Heather Viles

Professor Heather Viles was awarded the prestigious Founder's Medal by the Royal Geographical Society in 2020. She is a biogeomorphologist and was one of the first people to develop and champion this interesting element of physical geography. In an interview on *Women's Hour* in May 2020, Heather talked about the influence of her mum, who was also a geographer:

I became a geographer really because of my mum ... [who] studied geography and always encouraged me to look around in the world and ask questions ... [when] we used to go on family holidays... Mum always instilled in me this sense of enquiry about the world around us.[52]

Her early research was on the impact of acid rain and air pollution on cathedrals, and later work has ranged from a focus on corals in the Seychelles, to many of our planet's deserts, and more recently to helping conserve ancient archaeology from China's Silk Road.

Katie Hall

Dr Katie Hall works with esri UK to develop and deliver training courses and related materials for teachers and educators in a range of subjects, including geography. She has spent the last few years travelling the country, leading training for trainee teachers, beginners and those with a little more knowledge of GIS. Katie took geography at A level but then specialized in archaeology for her degree and postgraduate studies. She returned to geoarchaeology during her MPhil and PhD and strongly believes that geographers have some important skills, including

an appreciation for multiple different factors coming together to explain a situation – that simple answers are rarely correct. A willingness to balance 'human' and 'physical' to come to an understanding of the world and knowing the value of a good map as a tool for understanding.

She says that in her work with GIS, she combines her powerful subject knowledge with

lots of 'geography facts' that I draw on from memory every day to understand what I am looking at. Some of these I have had to update with more reading of recent findings and theories.

This serves as a reminder of the lifelong learning that geography 'forces' upon those who engage with it. Of her career path following her studies, she says:

Historical geography as part of my A level was what led me to study archaeology and set me on a path to completing a doctorate – something I would not have believed possible back in sixth form! Even if you don't stick with 'pure' geography, the tools of thought it delivers are helpful across a broad range of other careers.

Ben Hennig

Professor Ben Hennig is a geographer and cartographer currently working at Reykjavík University, having previously worked at the University of Oxford. He is the creator of the gridded cartograms that are at the heart of the Worldmapper website.

Ben's work has also appeared in national newspapers, and in journals such as *Geographical*. They're unmistakeable, featuring a world map that is both instantly recognizable and strangely shaped. Look at the cartogram on the opposite page, which was created by Ben. To understand it, you need to possess a certain level of geographical knowledge.

Annual precipitation (mm)

below 100 1500
250   1750
500   2000
750   2250
1000   2500
1250   above 2500

Figure 3 Worldmapper cartogram showing the global distribution of precipitation compared with population. (*Source*: image produced by Ben Hennig, and used with permission of Worldmapper.org.)

To interpret Figure 3, you need to know the usual position of the continents. If you know those things, the cartogram can help you appreciate which places have the highest annual precipitation (that is, which are the wettest places). The map shows countries proportional in size to their population. Annual precipitation is overlaid on it: lighter shaded areas receive less than 250 mm of rain a year, whereas the darkest areas receive more than 2,000 mm annually. It's possible to identify locations where water supply may be a particular issue by looking for lighter shading in areas with lots of people.

Ben travels the world making these maps. Of his own geography background, he says:

> In my everyday life it is probably inevitable to see, understand and judge everything that happens through 'geography-tinted' glasses. Once a geographer, you will always stay one and will find it hard to not utilize this way of thinking in your everyday life... Geospatial technology is something I quite enjoy integrating into my normal life even when it is not about research or teaching.
>
> In our day and age everything has become ever more specialized, so my recommendation for prospective students is to look at how geographers of the past understood the world around us with a much more holistic and universal perspective.

He told me that studying geography had given him a love of travel, all of which stemmed from his introduction to GIS. Ben continues:

> What really got me hooked in terms of an academic interest was the world of GIS, which I had not heard about at school before. Being able to learn GIS made me switch from geography as a minor (I started studying German language and literature as a major along with philosophy and geography) to undertaking a full geography degree

and a very interdisciplinary way of thinking, integrating both human and physical geography perspectives which I aim to still consider in everything that I am working on.

Ben's recent working includes tracking the impact of Covid-19 and producing animations of the changing locations of major outbreaks as they have shifted to different locations around the world. These animations are made up of lots of individual maps that Ben creates using large sets of data and a range of software tools (https://worldmapper.org/covid-19-coronavirus).

Ian Cook

Professor Ian Cook works in the geography department at Exeter University. Ian's interests are in the stories behind the commodities that we buy and the supply chains that bring them to us. He adds '*et al.*' to his name when publishing papers because of the collaborative nature of his work and of projects within geography more generally.

His research has been wide ranging but has broadly fallen within the fields of activism and commodities. If you are the sort of person who likes going on marches and protesting – perhaps you've been involved with Extinction Rebellion or other youth activism – Ian may be your kind of geographer. Check out his website 'Follow the Things': http://followthethings.com.

His early research involved following the story of a papaya from a supermarket in the UK back to the field in Jamaica where it was grown, unpicking the papaya's story at each stage of its journey, and telling the tales of the people who were employed in the chain. He has also used scenes constructed from Lego bricks to protest about injustice in garment factories in Bangladesh and about other political situations.[53]

In a film made for the Time for Geography website, Ian reminds those who might want to carry out geography research that when picking a topic, you should ask yourself whether you are actually interested in the topic: has it caught your imagination?[54]

Alan Kinder

Alan Kinder is the chief executive of the Geographical Association, a charity founded in 1893 by a small group of geography teachers with the mission to 'further geographical knowledge and understanding through education'.

In a video that can be viewed on the Geographical Association's website,[55] Alan describes the changes that have taken place in the school subject since the association was founded in 1893 as being

from a gazetteer, a way of trying to cover the world, towards a subject which seeks to explain and explore the world in which we live... There's never been a time when geography education has been more important than today.

Alan's career has included periods of time working as a teacher, as a tutor in a field studies centre, and as an advisory teacher in Barking and Dagenham in London.

Corinna Hawkes

Professor Corinna Hawkes is the director of the Centre for Food Policy at City, University of London. She gained a geography degree from Bristol University before going on to complete a PhD at King's College London. Corinna told me that she first became interested in physical geography at school, where she realized that:

There are forces shaping the world around us in very tangible ways, all the time. It's impossible to disconnect them. When one thing changes, it has a knock-on effect on something else. Try as we may to separate things out, we cannot. The shape of the world around us is the sum of the interaction between these forces at multiple scales.

After her PhD, she went on to work in food systems. This is a really important area as it has vital links with people's health and life expectancy. Her geographical thinking is vital here too:

I look at how policy and action can intervene in food systems to help the world eat better. From this I know that what we eat is connected to practically everything out there in the world; eating is a human activity affected by our ecosystems, society, economics and politics and the result of numerous global forces. What we eat in turn has multiple impacts on the world, not least climate change. For several years I studied globalization and trade and how this influences what we eat. We might look at our plates and think we have just chosen it, but it's the result of stuff that goes on far away that we cannot see. I am fascinated how what is on our plates every day is influenced by what happens far away, including economic and political forces that are hard to see.

She also believes that geography is vital in providing context for her current work:

From geography I learned that everything is connected. It is this skill which has led me to succeed in my chosen field. Employers are looking for people who understand how the world works and can make decisions in that context. Geography helps us understand how the world really works.

Kit Rackley

Kit Rackley has had an interesting and varied career and is currently a freelance geography educator. They studied environmental studies at the University of East Anglia, which provided them with both a way of looking at the world that has persisted ever since and an appreciation of issues surrounding sustainability. Kit worked as a geography teacher for some years in Norfolk before changing career for the first time. They are passionate about geography. Here is Kit reminiscing about first realizing they were into the subject:

I almost felt like geography was a super-power. I was learning about how the world works and could start seeing it with 'x-ray vision'. When I was eleven my family went on our usual beach holiday. I spent more time looking for clues into coastal processes and explaining to my parents what 'longshore drift' was than actually goofing around. From then on, I knew I was going to study the subject.

After 13 years working as a geography teacher, Kit moved to the US to work first at the National Oceanic and Atmospheric Administration in Boulder, Colorado, and then at the Exploratorium, a museum of science and technology and a popular visitor attraction, in San Francisco.

Kit told me:

Over the past few years I have been working with world-leading climate scientists and academics, taking complex scientific processes and information and communicating it to students and members of the public. I certainly did not understand a lot of the details that the scientists were working on, but my geographical knowledge was deep enough to understand the elements that were important for non-experts to know, and challenge misconceptions.

Kit later returned to the UK to work at the head office of the World Energy and Meteorology Council within the University of East Anglia. They took up a position as a project manager and education lead on the C3S Education tool mentioned in Chapter 2.

They say of geography:

Geography has opened doors for me that I never dreamed of, or never even thought about. I used to say to students thinking about studying geography related subjects that 'you'll never close any doors opting for geography but keep them open, and even open more you didn't realise were there'. Geography is so rich, diverse and interconnected that even if you specialize in one area, you'll be able to make connections to others. My job connected me to scientists, engineers, policymakers, educators, environmentalists, just to name a few. I ended up specializing in meteorology and climate change, something I never thought of when I first started my degree.

Lucy Verasamy

Lucy Verasamy works as a weather presenter for ITV News and Sport, having previously worked for MeteoGroup UK, Sky News and ITV's *Daybreak*.

Lucy studied A level geography at a school in Norfolk, where she had the great good fortune to have the author of this book as her geography teacher. She knew she wanted to be a weather presenter, so her next step after school was to do a geography degree. She chose to study with Professor Iain Stewart at Brunel University, going on from there to write forecasts for MeteoGroup UK and eventually (sometimes dreams do come true!) to become a regular face on our TV screens presenting the weather.

While the people discussed in this chapter have, perhaps, taken their geography studies to a particularly elevated level, there is no shortage of examples of geography graduates who have ended up in more everyday occupations. The following vignettes were provided by two Cardiff University geography graduates.

Nick's undergraduate degree was in human geography and planning, and he went on from that to complete a master's in international planning and development. He now works as a senior planner at Savills, the international estate agent.

Nick told me where his interest in geography came from:

My inspiration to do geography and planning at university came from a GCSE geography exam question on the social, physical, environmental and economic impacts of regeneration of land adjacent to the River Lagan in Belfast between 1989 and 2000. This interest in the built environment then led me to a career in planning. Being a town planner relies on the ability to read and interpret maps, whether they are showing land ownerships, constraints or opportunities, and geography provided me with the toolkit of skills to do this.

Anish graduated with an undergraduate degree in human geography in 2015. Since graduating he's worked at Barclays Bank as an associate director in the corporate investment banking team, specializing in real estate financing. Here's Anish's geography story:

When leaving school, it was difficult to decide what career path I wanted to pursue. I decided to study human geography at university as it was my favourite subject but importantly would provide me with the transferable knowledge and a skillset to apply to a wide array of societal issues and career opportunities. Working in banking and finance requires you to have a broad understanding of a range of topics and the associated socio-economic impact which could potentially impact any transactions that are being worked on. Having a sound grounding in topics such as geopolitics, the regeneration of cities and globalization has helped me to identify risks in transactions which other colleagues may have approached from a different perspective.

Some things to think about before reading the next chapter

▸ Having read the profiles of the people in this chapter, who do you think had the most interesting routes into geography? Whose career pathways were most intriguing to you? Can you think of further questions you would want to ask these people if you met them? And finally, which of their eventual/current careers do you think is most appealing?

▸ The Black Lives Matter movement grew in prominence following the tragic death of George Floyd in May 2020. Explore the extent to which the geographical community reflects the diversity of the population at large. What more could be done to improve the recognition of black geographers?

▸ Do some further reading on the value of geographical thinking: there is a world of books out there for you to dive into. Try to complete the following simple statement using everything you have gained from reading this book up to this point: 'Geography matters because...'. I'll take up this theme again in Chapter 7, but I hope you've already gained plenty of ideas about what to say!

CHAPTER 5

Preparing to study geography at university

Explore how the world around you works, and how people live within it.

— From the geography subject guide produced by UCAS

THIS CHAPTER WILL EXPLORE THE process of applying for a place on a UK university degree course. It will also explain why a geography degree is the one you should choose! It will identify important aspects of university life that you should anticipate, and perhaps start preparing for. Some of these are generic to all courses; others are specific to geography. I will provide advice for both students and parents. Note that while this chapter has a UK focus, there will be some areas of overlap with the processes in other countries.

This is a particularly interesting time to be considering university study, and the changing dynamics of the application process will have an influence on when people start their studies. While there have been few changes to the application process itself, the decision-making dynamic may have changed for many, and the use of more online learning and teaching might not be to everyone's taste. A 2020 report from the Sutton Trust notes that changes in the application process might affect students differently, and not always equally.

The first step when preparing yourself for study at university is to familiarize yourself with the UCAS process. More comprehensive resources than this book are available to those who want the full UCAS picture, but all you need to know for now is that if you make the decision to study geography at university, you will need to make use of the UCAS website both to explore the available courses and to begin the process of applying for a university place.

UCAS stands for the Universities and Colleges Admissions Service. Its website (https://ucas.com) has a number of subject guides, each of which provides extensive details about university courses in its subject. As you read through the geography guide, you will see that the subject is offered in numerous combinations, both as a single honours course and in combination with a wide range of other subjects.

You will need to select a number of potential university choices, and these will make up your 'UCAS track'. The whole process is now completed online. Another thing you'll need to investigate is the number of UCAS points, or the examination grades, that will usually be required if you want to access a particular course. You'll need to know this to ensure your choices are realistic. The admission criteria for some geography courses will be based on grades, whereas others will be based on points gained for exam success. Another resource here is the UCAS 'tariff calculator' (www.ucas.com/ucas/tariff-calculator).

UCAS regularly releases applications data, and its June 2020 report included the encouraging statistic that applications for geography degrees were up on 2019. This is part of a wider trend that saw an increase of 1.4% in total applications for university places.

Your first step should be to go to the UCAS website and create a personal UCAS 'hub', where you can search for courses, keep notes and build your personal statement. The 'Unibuddy' part of the site will give you the chance to speak to students who are already studying geography and ask them questions.

You will need to complete a UCAS form and write a personal statement as part of the process. This is all done online. You'll also need to provide references: these are usually written by your subject teachers or form tutors with the support of pastoral staff within your school or college.

It's a good idea to make early contact with the staff at your school or college whose job it is to guide you through the process. They will have lots of experience in supporting students and providing the best advice. They may also have a step-by-step guide that you can follow. You'll also need to make yourself aware of the important dates by which certain stages of the process need to be completed. There's always an element of flexibility, and, although it's not recommended, a decision to apply can even be made in the summer term before the course starts. This is done via UCAS Extra if it's before 30 June, or you could carry out some research in preparation for the clearing process. Reduce your stress levels by applying early, and exploit any and all available guidance from teachers.

What requirements are there for studying geography at undergraduate level?

The UCAS subject guide makes the following general points about applying to study geography at university.[56]

A levels

Entry requirements range from CCC to AAA, with universities and colleges most commonly asking for BBB.

Scottish Highers

Entry requirements for Highers (the most common qualification in Scotland) range from BBBB to AAABB, with universities or colleges most frequently requiring AABBB. Universities occasionally ask for Advanced Highers to supplement Highers. If Advanced Highers are requested, universities or colleges typically ask for AB.

Vocational courses

Other Level 3/Level 6 qualifications (e.g. the Pearson BTEC Level 3 National Extended Diploma, or an SCQF Level 6) may be accepted as an alternative to A levels/Highers by some providers. It's essential that you check alternative entry requirements with universities or colleges.

There may also be separate admissions requirements for individual colleges based on their individual preferences or interviews. Note that the overall demand for particular subjects in particular years, alongside the capacity to take a particular number of students, may also have a bearing on offers and so on. You should be able to find a course that suits your interests and qualification level, as the range is very wide. Be ambitious rather than conservative with your choices. If you have a keen interest in the subject, you are more likely to be of interest to those offering places on courses. Check that your school grades meet or exceed the entry requirements to avoid an instant rejection.

It's also worth checking whether other qualifications you may have taken, e.g. LAMDA (for drama) or ABRSM (for music), will count towards your tariff using the calculator we mentioned earlier.

Which A levels should I take in order to study geography at university?

There are no particular subject requirements for acceptance onto most degree courses, but having studied geography at school is probably a good idea, of course, if you want to do a geography degree. Questions will certainly be asked if it's not on the list of A levels you took, although this won't necessarily cost you a place. I applied for geography degree courses having studied English, mathematics and computer science along with my geography A level and 'general studies', but others start their undergraduate geography studies with a more creative mix of subjects or with a grounding in the sciences. Some courses have a strong focus on physical or atmospheric systems, where a strength in mathematics or physics is likely to come in handy, and different institutions have different requirements when it comes to mathematics grades (even going back to GCSE (or equivalent)). I took degree modules in statistical analysis and in computer mapping, and while they were both more related to geographical skills than to maths, they did have a mathematical aspect to them that my A level maths background certainly helped with.

It's important to note that geography A level is useful for many more students than just those who want to do a geography degree: it is widely considered an acceptable subject for courses in law and medicine, for instance. An all-science route isn't the only way to access those courses. Consult with experts, as there are some misconceptions surrounding geography here.

Extended Project Qualifications

Some post-16 students complete an additional one-year course called an Extended Project Qualification (EPQ). This can be a great way to connect your other studies with geography.

EPQs provide you with a relatively free choice of topic to study and research: in effect, it is an extended essay with referencing. They offer an excellent opportunity to further develop areas of geographical interest. Students need to choose a topic area to study that is linked to the subject but that is not already covered by the specification. It has to be something 'new'. The dynamic and synoptic nature of the subject allows for great flexibility in topics and offers the chance to extend your passion for the subject while improving your research skills. I have seen students exploring academic geography critically, investigating change in remote locations and considering influences on their own personal geographies.

A 2016 report found that around 40,000 students took this qualification. Around a quarter of them went on to do a mathematics degree, but just over 10% of the candidates chose to study geography. Completing an EPQ provides a number of benefits if you are considering undergraduate study in geography.

► Choosing a particular geographical topic for your EPQ may allow you the chance to research a particular area of the subject.

▶ EPQs provide you with an opportunity to develop your ability to use Harvard referencing, or to familiarize yourself with other referencing systems and styles of academic writing that you will need to get to grips with later. You might read geographical books pitched at a slightly higher level than would otherwise have been the case, expanding your geographical vocabulary.

▶ EPQ studies may provide you with a suitable topic for discussion in an interview for a university place. It's good to have something that you will feel comfortable discussing, having spent some considerable time exploring the topic.

▶ It will be useful when writing your personal statement to show your commitment to more focused study as well as highlighting your broader qualifications base. This can be done by outlining the topic you are studying for your EPQ and discussing its geographical relevance.

▶ Universities offer UCAS points for completion of an EPQ that are equivalent to half an A level. This could make all the difference when it comes to being accepted onto a course.

▶ Universities may also give 'contextual' offers to students with good predicted grades in their EPQs alongside other predicted grades.

The status of geography in higher education has grown in recent decades. We have previously mentioned **STEM**, which refers to discipines linked to **s**cience, **t**echnology, **e**ngineering and **m**athematics. These subjects have been the focus of a great deal of attention over the last decade, and they have received substantial additional government funding, as they are seen as essential to the economic development of the country. The RGS-IBG has been active in raising the status of geography as a contributor to STEM provision, and geography is now formally recognized by the Higher Education Funding Council for England as 'part-STEM' in higher education. This has raised the status of the subject, and geography is unique in that it's a humanity, a social science and has this part-STEM designation. It's also addictive.

In June 2020, a new initiative called **SHAPE** was launched by the London School of Economics and Political Science, the British Academy and Arts Council England (https://thisisshape.org.uk/). The initiative proposes that alongside STEM should sit the 'social sciences, humanities and the arts for people and the environment' (SHAPE). We will see in time whether this has the same traction as STEM, but the participation of organizations such as the RGS-IBG should help the idea take hold as it develops and takes shape. It is certainly a reminder of the importance of the humanities: as the associated website says, 'SHAPE subjects teach us to understand more profoundly the world around us and the people in it'. That sounds to me a lot like a need for geography.

Because the skills learned during geography A level are support-ive of work carried out at university level, and because the subject combines well with others, A level geography was, until May 2019, included on a list of **facilitating subjects** that was shared by Russell Group universities. The study of geography was said to support entry into higher education. The Russell Group is a group of 24 (self-selecting) universities that have a particular focus on research. More details about them can be found on their website.[57]

The Russell Group's facilitating subject list is no longer used because it was felt that more weight needed to be given to arts and technical subjects, but the value of geography certainly remains unchanged. The list was replaced by the Informed Subjects website, where the latest information can now be viewed.[58]

Which geography degree is right for you?

There are hundreds of options for those wanting to study geography at undergraduate level. A search on the UCAS portal in May 2020 revealed details of more than 700 geography-related courses being offered by almost 100 different providers. Geography can be studied in its own right – as a BA (Hons) or BSc (Hons), or an MA or MSc (an additional option for those studying at universities such as Oxford and Cambridge) – and it combines very well with a wide range of other subjects, including (but not limited to) anthropology, archaeology, biology, business management, economics/economic history, education, environmental mathematics, geology, innovation, management, oceanography and planning.

It is well worth considering whether you want to combine geography with another subject: studying subjects in tandem can enhance both. You can search for degree courses on the Discover-Uni website (https://discoveruni.gov.uk) to find out more specific details about particular courses.

Something we can't avoid mentioning is that some universities are more prestigious than others, and this can influence students' future success. People who take classics degrees in Greek (which one might think would have limited employment potential) tend, like geographers, to be highly paid, but it is not the practicality of the degree that gets them the highly paid job: it is the university they went to and certain geographical biases in job selection.

Another example of this phenomenon is the fact that the majority of candidates to be Prime Minister after Theresa May resigned in July 2019 had degrees from Oxford, as have many members of parliament. It may be worth being ambitious with your preferences, but take advice from your teachers: there are, after all, a limited number of spaces on your UCAS track.

Note, too, that research into the graduate market by High Fliers (www.highfliers.co.uk) has found that work experience is more strongly correlated with recruitment success than where a candidate studied.

This may also be the time for some **scenario planning**. Try to visualize yourself in the future – the near future if you are already studying at university. Try asking yourself the classic interview question: 'Where do you see yourself in five years' time?' What job could you have that you would be happy to get up for and commute to each morning?

League tables are produced for all university degree subjects. They take into account numerous factors, including student satisfaction, graduate employment, entry requirements and the quality of the research being carried out in a department (which is measured externally). The tables can be informative, but you should look at them knowing that they are quite subjective, and don't tell the whole story. While I'd generally recommend treating them with scepticism, then, a high positive factor can still give you confidence in a particular course, or university. As a student, the quality of teaching you are going to receive will obviously be a particular priority, and you will certainly want to visit any university you plan to spend three years of your life at in person before making a final decision.

In 2020, the University of St Andrews (where, as you might remember, Prince William studied geography) topped a *Guardian* newspaper table of best universities, with the University of Oxford in second place and Durham in third. The *Complete Universities Guide*, by contrast, placed Cambridge in first place, so you may benefit from comparing different tables.

Teaching at universities is also graded according to the **Teaching Excellence Framework**, which gives grades of bronze, silver or gold.

You can check out the current rating of your chosen institutions on the Office for Students website.[59]

Many students will consider the quality of teaching alongside things like the number of contact hours with tutors as well as the actual course content. It really does pay to do your research in advance of making your final choices, so here are a few more suggestions for preparing to submit your application.

Read **geography departments' web pages**. You'll find a range of information about staff research projects and publications there, and it may well be more revealing than the more polished outlines of teaching content you find in prospectuses. All geography degree courses will cover the core concepts and methods of analysis, but the teaching will be strongly influenced by the research expertise of individual staff members; this means that the precise topics, techniques and geographical areas covered will vary considerably between institutions. Pick a course that matches your own interests and piques your curiosity. Department web pages usually also have a news section: use this to see how active a department is, and what exciting research projects have recently been funded. These may give you an opportunity to travel and develop your experience, giving you employable skills in the process.

Think about whether a **research-intensive university** or a **teaching-intensive university** is best for you. The particular focus and interests of an institution's academic staff are likely to have a bearing on the nature of teaching and on its relative importance in courses. Teaching-intensive courses will give you more contact hours, and you'll probably find that staff are more readily available to help students with coursework (because they are spending less time on research). Research-intensive courses may expose you to some of the cutting-edge research and analysis that is being carried out, but students on these courses need to be quite independent and self-motivated: you may find that your lecturer is on a field course when you need them, or that they're unavailable when you need support.

How is the teaching delivered? Some universities, such as Cambridge and Oxford, have a seminar system, with small group discussions and extensive preparation and later analysis. You may prefer lectures or types of teaching. Find out which courses do more fieldwork and laboratory-based work if this is your preferred way of learning. Prior to 1992, a number of polytechnics provided slightly more 'vocational' courses, with a stronger STEM focus, and it is still possible to find such courses in the universities that those institutions became. Do you want to spend more time outside in the landscape, investigating practical and applied aspects of the subject rather than more theoretical aspects?

The best route into a good university is to secure good grades. This will come from developing good habits of mind and application: something that has to start much lower down the school than year 13. Some universities, in some years, have **more available places than applicants**. When that happens they may be flexible on entry offers/tariffs if they are persuaded that an applicant is well motivated and has the potential to develop once they start a course. (English and mathematics GCSE grades are sometimes looked at to judge a student's potential if their A level grades are below par, so be sure to focus on getting the highest possible grades in those subjects in particular if you are still studying or resitting them.)

Ensure that your **personal statement** is coherent and grammatically correct. Personal statements are particularly important if you haven't studied A level geography, if your predicted grades are below the standard entry tariff, or if you're applying to one of the small number of 'elite' universities that needs to narrow down the 'best' potential students from a large cohort of applicants. Those institutions will read the statements of all applicants as part of the selection process. Statements become particularly important on results day if you decide to enter the clearing process. They provide you with a chance to introduce yourself to potential lecturers, so be sure to tell the story of your geographical journey so far, discussing the steps you have taken to broaden your horizons and experience the world. Make the most of any important geographical experiences. Try to avoid being 'passionate' about everything: it's a bit of a cliché. Ask others to read your statement as well: parents, for example, should be encouraged to offer constructive criticism. We'll suggest more guidance about personal statements later in this chapter.

Students enjoy studying geography

Every year, undergraduate students are asked how happy they are with their higher education courses. This process is called the National Student Satisfaction (NSS) survey, and these NSS scores are important to many when they are selecting a course or subject.

Geography students who are on a course with a high satisfaction score are more likely to complete their course, and less likely to drop out. Given the cost of completing a degree, with high tuition fees and living expenses, this is very important: you wouldn't want to start a degree course only to drop out part of the way through (potentially owing thousands of pounds to the Student Loan Company at the time).

It's important to note that geography has the third-lowest drop-out rate of any subject studied at university. This makes it a great choice for those who are intending to stay the course.

The 2020 NSS survey, which was completed by hundreds of final-year students at every UK university, revealed that 88% of the respondents studying geography were satisfied with the overall quality of their course. That's higher than the 83% nationwide average for all respondents and all courses.

The survey also indicated that geography students were highly satisfied with specific elements of their university experience: the organization and smooth running of their courses and their ability to access specialist resources and facilities were frequently mentioned. All of this is important when you are committing a period of your life to the sustained study of a certain discipline.

You can look at the latest graduate destinations on the DiscoverUni website (https://discoveruni.gov.uk), and further details on the NSS survey itself can be found on the Office for Students website.[60] The Graduate Outcomes site (www.graduateoutcomes.ac.uk) is another good source of material that reinforces the idea that geography graduates have many and varied talents.

Another source of relevant information here is the Quality Assurance Agency (QAA): an independent UK body entrusted with monitoring and advising on standards and quality in UK higher education. It helps to ensure that the three million students who are working towards a UK qualification at any one time get the experience they are entitled to expect. It works across the UK, and part of its remit is to help build international partnerships that enhance and promote the reputation of UK higher education worldwide. The QAA also produces 'subject benchmark statements'. These are written by subject specialists and they outline the nature of study in a subject as well as detailing the academic standards expected of graduates: they discuss what graduates might reasonably be expected to know, do and understand at the end of their studies. These statements are used as reference points in the design, delivery and review of courses. They are produced for bachelor's degrees at honours or master's level, and for professional qualifications in Scotland.

Another way to find out more about a geography degree course that you are thinking of applying for is to see whether it's been accredited by the RGS-IBG: the accreditation process guarantees quality against certain standards that have been identified by the RGS-IBG. You may find it helpful to check if the university course you are interested in has gained this accreditation as an extra step in your 'due diligence'.[61]

Choosing a *place* at which to study geography

When it comes to choosing a place to study, it's perhaps time to apply a little geographical theory to your decision. Hull, Leeds and Exeter are all places, obviously. They are marked on maps; you can visit them in person. However, when geographers talk about place, you will remember that they also consider the *concept* of place.

Places may have a reputation, for example. When it comes to university degrees, some places may be more 'impressive' than others to those who know nothing of the actual practicalities of study. If you asked a layperson to name a top university, they are likely to mention Oxford and/or Cambridge, but other universities also rate highly when it comes to student satisfaction and other indicators of quality.

Places are also described as being 'socially constructed'. This aspect is of particular interest to those students whose social life is important. The identity of some cities comes, at least partly, from the presence of large numbers of students. The spending power of those students can be significant too. Demographics influence the presence or absence of certain services and retail premises, as will the percentage of certain nationalities that live in the residential areas close to the university campus where you are likely to live and shop.

Places are also affected by a number of internal (endogenous) and external (exogenous) factors. The best way to get a feel for whether you want to spend a lengthy period of time in an unfamiliar place is to visit it in person. Try to be there at different times of day, too: a visit for a few hours on a Wednesday afternoon may be very different to going on a busy Friday night; staying overnight might make you realize that the seagulls are really loud and will affect your sleep, or that the sun rising over the sea is balm for the soul. Think about how students are viewed by locals, or about what part-time job opportunities might be available: perhaps you could become one of the students in a straw hat and waistcoat who frequently pester visitors to Cambridge by asking if they want to go punting on the River Cam? Find out about the general cost of living. Investigate the options for self-catering in halls of residence or for going full-board. I remember one university housemate who lived on Pot Noodles for a year and another who used to cook on a wok in the garden on a fire made from branches snapped off trees on the way home from daily lectures (although this was in the days before Deliveroo).

One thing's for certain: you will always remember your university days. They'll provide you with (mostly good) lifelong memories and strong friendships – even life partners – but try to avoid getting a dodgy tattoo. For some, who find an affinity with the city or town they have chosen to study in, it becomes the place where they continue to live and work once their studies are over. Geographers call this 'place attachment', but shepherds have a nicer word for something similar: the term 'hefted' is used to refer to sheep who have become so used to living in a particular area that they seldom stray from it even when fences are removed. Choosing a university may therefore mean choosing your future home city.

Making the most of a geography open day

Open days are a really important part of the process of choosing a university course. Once the calendar for these is published, visit as many as is feasible. If you can't make it in person, some may offer virtual tours, but the experience will be no replacement for actually visiting a place. Visit the geography department, of course, and speak to some of the lecturers who you might be taught by. It's a great opportunity to ask the questions you want answers to. It is always worth planning ahead and booking a place, as open days tend to fill up. The UCAS website is a good source of guidance on open days.

As you walk around a campus, think about the daily experiences you might have if you were studying at this university. Buy a local paper and flick through it, perhaps. Is the news that's being reported in its pages mostly about crime, or about local successes and opportunities? Can you find good coffee shops? Are the kinds of cultural opportunities that are important to you available?

Also consider how easy it is to travel to the campus, both from your home and from the university accommodation that is offered. How expensive will these journeys be if you are travelling home more frequently than once or twice a term, say? If you won't be driving, consider the location of train stations and the frequency of services. Depending on which country you live in, some of the distances involved in these journeys could be very large.

Check out the facilities you're going to be using regularly within the geography department building. Are they modern and up to date? Are lectures mostly in one area of the campus, or will you be travelling all over? Are lecture theatres comfortable and well equipped? Are the library facilities cramped and oversubscribed or does it seem like you'll be able to find a quiet space to study? How good is the campus wi-fi? How large are the study groups? How many hours of contact time will you have with academic staff? What support services are in place should issues arise? Are there specific needs you have that are better met at a particular university, with someone who can sit down with you and talk through your concerns?

Visit the library and explore the geography section of the book stacks. Are there multiple copies of key books in the catalogue? What digital resources will you be given access to? Which journals are available to read?

Consider the social side of campus life as well. Visit the social spaces and sporting facilities if these are important to you.

Entertainment may also be part of your interests: if so, are there cultural venues nearby? What are the options for clubs and societies with a geographical connection, such as climbing?

Visit the student accommodation. What is involved in applying for this? Is there sufficient, modern student accommodation for all first-year students who request it? Are there shops on campus that sell the essentials you'll need? How expensive is it to live in your chosen city? How much is a pint of beer?

If you really can't make it along in person, find out about virtual visits. These have improved in recent years, with increased use of virtual reality: many institutions have used drones and 360° cameras to recreate the experience of visiting their campus. Some have had entirely online open days in 2020, with virtual tours and sample lectures and talks that can be viewed on the university's website. Remember that many geography departments use drones in tuition and fieldwork too, so you may even come out of your studies with a pilot's licence.

As well as open days, keep an eye out for 'taster days' or 'insight days' for geography-related courses and sign up if you can. These are usually put on between Easter and the summer and can be particularly helpful for year 12 students. You can meet other like-minded geographers at these events, and they might give you something to add to your personal statement.

Academics and social media

It's worth noting that most university geography departments have active Twitter feeds, as do many individual academic staff. The feeds can be a good place to find information about the work and research being carried out in a department. You might like the particular tone of certain accounts. If so, find out more about that person or department. Perhaps you'll be interested enough to respond to one of their tweets with a question, forming a direct line of communication. Some example accounts are mentioned in the references section at the end of the book, but it's not an exhaustive list, and academics frequently move to different institutions, too.

Applying for geography degrees

This book won't walk you through the whole UCAS process – the UCAS website and other dedicated publications are available to help you with that. I can tell you that the system has been helpfully simplified in recent years.

Some universities are particularly interested in recruiting students who have high 'cultural capital': those who can talk knowledgeably about the subject they're applying for and place it in a wider context. You can develop this skill not only by studying the academic textbooks that your teachers may ask you to use, but also by reading more widely and even by tracking social media accounts and reading a daily newspaper.

Application procedures vary a little between universities and colleges, so do your homework in this area too. University prospectuses will generally provide guidance on deadlines and expectations. See whether you can connect with former students at your school who have studied geography at the particular universities you're interested in. Your teachers will also know of people who went on to study geography at university who may be able to provide some recommendations for you, as your school will track the destinations of leavers.

Writing a personal statement

As mentioned earlier in the chapter, you'll be asked to write a personal statement as part of the UCAS process. Once written, you should try to have it checked by a teacher before you upload it to your UCAS 'hub' area. You should also identify referees who are willing to write about you. One of them is likely to be your geography teacher. Your headteacher will be supported by your geography teacher in creating your final reference, as they will be able to assess your potential in geography better than other colleagues. Cultivate and nurture the relationships you have with your teachers. It's also worth remembering that the reference will probably mention your level of attendance and your attitude to learning, so make sure you make it into school every day unless you're really sick, and avoid getting detentions!

Personal statements are limited to 4,000 characters (including spaces!) or 47 lines, whichever you reach first: this is a challenging restriction that requires you to write concisely. The UCAS website provides guidance on how to use your allotted words carefully. It's hard to say everything you might want to within that character limit!

Talk about your interests from the first line, and make sure it's clear that your interest is with geography. Rather than saying 'I love geography', though, you should probably choose specific areas of the subject that particularly interest you. One good recent example that I've seen was: 'After a visit to the Goldsmith Street area of Norwich with school, I am particularly interested in how sustainable housing can make a real difference to the homeowner's quality of life.' You also need to stick to the point. One approach would be to paraphrase Michael Palin and write a variation on what he wrote about geography in 2011:

Geography is a living, breathing subject constantly adapting itself to change. It is dynamic and relevant, a great adventure with purpose.

You'd be better off, though, trying to demonstrate your enthusiasm for the subject by drawing on the additional experiences you have had. Why *do* you want to study geography? Try to show how you have connected what you've learned on your geography course to your life and future plans. Avoid the word 'passion', perhaps: reserve that for people, or fruit.

Explain why you're interested in studying the particular course you've applied for. Talk about any additional evidence of your dedication to geography (as I discussed in Chapter 2): the fieldtrip you took, the work experience you arranged, the guest speaker who came to your school, etc. This may help you stand out from other candidates, which is what's required. One Cambridge lecturer said of Tim Marshall's *Prisoners of Geography* that, according to the personal statements she read, it must have been on the bookshelf of everyone who'd applied for her course. While this might have been said with tongue firmly in cheek, it is another reminder to try to read beyond the best-sellers list.

Ensure you take a healthy interest in the environment, politics (particularly beyond the UK) and the world around you. You should certainly have, and be able to demonstrate, a curiosity about not only the changes themselves but the processes that have led to them. Mention any experience you have of the world of work – even if it's a part-time job or some brief work experience – that has a geographical theme.

Here are a few more pointers.

▶ Be aware of some of the key concepts of geography and be able to talk knowledgeably about them. Lee and Dorling's book *Geography*, mentioned in Chapter 8, is a particularly useful resource for this section of the form – and it can easily be read in half a day, so it's very accessible.

▶ Take advice from the teacher in your school whose role it is to help support university entry. They might be able to give you some advice about specific universities that may suit your particular interests or strengths.

▶ The Field Studies Council runs courses to develop fieldwork skills. These are specifically designed for those students who are thinking of studying geography at university, so participation in them would be a good way of showing a certain level of commitment to the subject.

▶ There is a video on the Curious Geographer YouTube channel about how to put your statement together that is very helpful.[62] The channel is a good source of advice and useful resources more generally, too.

▶ Jane Marshall provides advice on behalf of UCAS, and her YouTube video on writing a good personal statement would also be a good thing for you to watch.[63]

It may be worth taking a Massively Open Online Course, or MOOC. These are often free, and having the term MOOC on your personal statement will help it stand out. Courses cover a wide range of topics, including globalization and supply chains.

One course provider is FutureLearn. They provide a range of free, certificated online courses: you can even obtain a paper certificate with the payment of a fee. Some of them have particular enrolment dates and run several times a year; others are available all, or most, of the time. One example worth checking out is 'Come rain or shine: understanding the weather', run by the University of Reading and the Royal Meteorological Society.[64]

Preparing for an interview for a geography course

You may be asked to interview for a place as part of the application process. While this has become less common in recent years, it's worth considering how you might improve your chances of being viewed favourably in this unfamiliar context. Start jotting down things you'd like to say. Interviews are more often used by prestigious universities, as they are keen to see first-hand the thought processes and creative thinking that potential undergraduates can show when placed under pressure. Interviews generally start with gentle questions to help you settle down, becoming more challenging as they progress.

If you have chosen to explore a geographical topic as an EPQ or have completed a MOOC, be ready to explain how it helped with your understanding of geography, and how you might develop or revisit it with additional geographical knowledge gained from further study.

Do some research into the university, its research interests and the specialisms of the department. What are the department's strengths? What key, funded research projects are currently live? Some universities have particular institutes on campus: Bristol University has the Cabot Institute, for example, and the University of Hull has the Energy and Environment Institute. The latter has been involved in a range of exciting games-based projects that connect research into flooding in the UK with the development of simulations into flooding, helping people prepare themselves appropriately.

You may be given an image to look at and be asked to talk about the geography within it. It could be a postcard, a map, an aerial photograph or some variation on this theme. Geography develops a skill called graphicacy, which involves analysing visual images. Practice this skill using striking images from daily newspapers or from the BBC's homepage. Try to become fluent in identifying important elements and coming up with key questions about any image. Who took this image and for what purpose? What is just beyond the edge of the image? Is everything quite as it seems at first glance?

Another interview possibility is that you'll be handed an artefact of some kind. This could be a rock sample – you might be asked to explain where you think it came from, or how it was formed, or what processes have affected it. This is perhaps more likely for degree courses that have a geological element to them. Always have in mind the difference between spatial and temporal elements: geographical processes happen over different timescales as well as at different scales.

You might be asked to respond to a scenario where there isn't really an incorrect answer, but which geographers would understand as contested. An example would be:

I was in the supermarket buying food and I needed some beans. I picked them up, but when I saw they were grown in Kenya I put them back because of the food miles. Did I do the right thing?

Although the exact process varies from university to university, the following are actual interview questions that have been asked. Consider how you'd answer them, and remember that in the actual interview you won't have very much thinking time.

- What do you most enjoy about geography and why?
- What do you think is the greatest geographical challenge facing the planet and why? What is the most complex global process? Which do you think is the most important of the Sustainable Development Goals?
- Which is easier to predict: climate change or future migration patterns?
- Can you explain sand dune succession? Or the development of river or coastal landforms? Or the growth of a village or town over time? (An interviewer might ask about any number of common geographical processes such as these.)
- Look out of the window and tell me about the geographical processes that you can see happening.

- An interviewer might point to a map on the wall, or perhaps show you a Worldmapper cartogram (like the one on page 98), and ask you what it shows.
- Explain what geography means to you.
- Explain why the price of oil is such a critical geopolitical construct.
- What is the different between 'place' and 'space' in geography?
- Describe how/why/when you think geography may cover new subject areas in the future.

Preparing for university study

Once you have a confirmed place, start looking at the modules you'll be taking in the first year. A reading list will be sent to you ahead of starting your course, and some of your lecturers may also provide material (in advance) in digital formats. In May 2020, James Cheshire of UCL's geography department shared an excellent video on the use of maps during a pandemic. It provided a useful introduction to the thinking that academics apply to particular contexts, and it can be viewed by anyone on YouTube.[65]

You might want to consider taking some free courses to upskill yourself. The Geographical Association has collaborated with the Open University to curate a list of suitable courses to get you started, all of which can be found on the association's website: see www.geography.org.uk/Open-University-resources.

It may be worth your while signing up for a free JSTOR account at www.jstor.org. This will provide you with access to a limited number of academic articles each month. This will give you a way of starting to explore the varied scholarship of geographers that has been written and shared. Should you go on to study at university, you will certainly make use of repositories of knowledge like this, and being able to navigate them is a very useful skill to acquire.

Some things to think about before reading the next chapter

▶ Do some research into universities that you might consider applying to. Use the websites suggested in this chapter to firm up a shortlist.

▶ Consider completing one or two of the online courses discussed in this chapter. If your university place has been confirmed, you'll probably have received a copy of the suggested reading list for your course, so get started on some reading.

▶ The UniGuide (formerly the *Which Guide*) has a subject-specific area that you could take a look at (www.theuniguide.co.uk/sub jects/geography). It provides guidance on appropriate courses that you may find useful.

CHAPTER 6

Making the most of your time at university

Anyone who leaves school without geography as part of their education cannot be considered to be fully educated.

— James Fairgrieve, geographer, educator, and
geopolitician, and president of the Geographical
Association in 1935 (speaking in 1925)

IN THIS CHAPTER I'LL PROVIDE some advice on how you might make the most of your time at university to prepare for what comes next. Whether you're already part of the way through your degree course or are about to start, you should be looking to explore all the available opportunities to make the most of your time at university. Note, too, that the city within which your university is located will provide additional geographical opportunities beyond the campus. You may never have this freedom and flexibility again, so seek out and embrace any and all possibilities for broadening your experiences.

Making the transition from school to university

Your first year at university is often a transition year, in which you'll gain the particular skills you'll need for undergraduate study. Lecturers will teach you how to apply your new knowledge and they'll introduce you to techniques of study, referencing, essay writing and use of source material. As well as learning to find your way in a potentially unfamiliar city, you'll also need to navigate your way through a complex subject landscape. Try to be as thorough as possible in your notetaking and ensure you keep on top of tasks. Technology can help, with many to-do list apps (such as Trello) available for your smartphone. Alternatively, you can simply use your phone's calendar app to help you keep organized, tracking meetings and other commitments.

Consider joining clubs and societies, and look out for other opportunities to be introduced to like-minded people or develop useful life skills, as well as practical ones.

It is never too early to start connecting your experiences with the academic theories you'll be exposed to. Be critical about information you're presented with, whether it is in the media, in books you're reading, or in articles in academic journals. Look out for hidden agendas: newspapers tend to have a particular political slant, for example.

If you're in the UK, consider sending an essay to the newly formed *Routes Journal*. This has been specially created to publicize the excellent geography being carried out by sixth-form and undergraduate geographers. The journal has a peer review system similar to those carried out by academic journals before a paper is accepted for publication, and it has set up a network of teachers to assess the quality of submissions. It offers a forum for all areas of geographical scholarship, including both human geography and physical geography. It is free and 'open access', with no paywalls or subscriptions. It could be a useful showcase for a piece of work you've completed in sixth form or as an undergraduate: seeing your work in print will always be a thrill. Check out its website, where you'll find details of how to submit your work: https://routesjournal.org.

What skills will you develop during your geography degree?

The UK's QAA sets the standards for all university programmes of study. It has produced a list of skills that are developed through a degree in geographical study, in addition to the specific geographical skills that are likely to be part of your course. Which items on the following checklist do you already have experience of from your A level or undergraduate studies? Try to find out more about any that you haven't yet experienced.

- Spatial awareness and observation.
- Abstraction and synthesis of information.
- Developing a reasoned argument.
- Assessing the merits of contrasting theories and explanations.
- Numeracy and statistical literacy.
- Handling large datasets.
- Preparing effective maps, diagrams and visualizations.
- Generating, collecting and recording primary data, and using secondary datasets (both quantitative and qualitative).

- Critically evaluating, interpreting and combining different types of geographical evidence (e.g. texts, imagery, archival data, maps, digitized and laboratory data).
- Analysis and problem-solving through quantitative and qualitative methods.
- Planning, designing and executing a piece of rigorous research or enquiry, both independently and in groups, including the production of a final report.
- Conducting fieldwork and field data collection.
- Employing a variety of interpretative methods (e.g. participant observation, ethnographic interviews, auto-ethnography).
- Employing a variety of social survey methods (e.g. questionnaire surveys and structured interviews).
- Employing a variety of science laboratory skills and methods (e.g. soil, water and plant sample preparation, microscopy, particle size analysis, soil and water chemical analysis).
- Collecting and analysing spatial and environmental information using a variety of methods (e.g. GIS, remote sensing, statistical and mathematical modelling).
- Taking responsibility for learning, and reflecting upon that learning.
- Recognizing the moral, ethical and safety issues involved in all aspects of geographical enquiry.
- Understanding the appropriate and ethical use of evidence and data.

The QAA's Geography Benchmark Statement was last updated in December 2019. It starts with this introductory statement:

Geography occupies a distinctive place in the world of learning, offering an integrated study of the complex reciprocal relationships between human societies and the physical, chemical and biological components of the Earth. Geographers study place, space and time, recognising the great differences and dynamics in cultures, political systems, economies, landscapes and environments across the world, and the links between them.[66]

A valuable characteristic of geography is its plurality of ways of knowing and understanding the world. The depth to which individual specialisms can be accommodated within its study is also highly valued.

One thing to look out for is whether there are opportunities for paid internships or placements – or, more likely, 'co op placements' – during your degree. These may last a term or longer in the later stages of a degree. HSBC Bank offers 'co op' placements, and these may be of interest to anyone thinking of a career in finance, for example. Universities have careers services that can offer guidance on these. If you intend to use these services, don't turn up with a vague request: target your questions around what you specifically need to find out that Google hasn't been able to help you with.

Some universities offer additional provision to students to help them become 'career-ready graduates'. Derby University, for example, has a professional development programme called STEPS: an acronym that suggests all graduates should be **s**killed, **t**echnical, **e**xpert and **p**rofessional, and that they should **s**tand out. The programme assigns graduates a personal tutor who can help them leave their course well prepared to move into industry or further study. Whatever is available to you in your own context, be sure to find out what it is and take advantage of any opportunities offered.

One final benefit of university is that, for many, it provides an opportunity to live away from home for the first time. Now's your chance to develop the basic skills that you hadn't previously realized you didn't have: how exactly do you iron a shirt, cook a poached egg with a runny yolk, or make sure you get up for an early lecture without a shouted reminder from a parent to 'get out of bed'?

Remember to keep asking the 'big' questions as well. Reading shouldn't be a passive process any more. Ask youself who wrote a particular piece. What was their motivation? Do you agree with what they're saying? Are there any hidden biases? Debate the issues; the more you read, the more you'll be equipped with alternative viewpoints. Draw on the knowledge of your lecturers here, and note that studying abroad will give you further perspectives on all of this too.

Compulsory modules

There are usually some modules that are compulsory. This is to ensure a degree of comparability between students early on in the course, and to teach some of geography's key underpinning ideas and outline how disciplinary thought developed. This helps you see where the subject originally 'came from' and how it has changed. These modules will generally help students make the transition from school thinking to university thinking. If you have been well prepared by your schoolteachers, and especially if you've also done additional reading for yourself, you may find these slightly easier and more accessible than others do.

Academic writing is also rather different to the type of writing you may have done previously. For example, it's unlikely you'll have been asked to write several thousand words very often before. This is made harder by the need for each statement that you make to be backed up with a reference ('Parkinson 2020', for example), and the requirement that every relevant article or book be added to a growing list of references at the end of each of your essays. Should you find yourself writing too much, you'll then have to edit your work down or you might end up being penalized. Writing can be hard work!

Optional modules

Optional modules offer you a chance to specialize. These often come in during the second or third year of a degree course, and they're likely to draw on the research interests of staff, and will therefore vary widely from university to university. This is where your choices count. Look at the available modules. Remember that if a lecturer moves on during a course, their module may need to be replaced, so don't get too hung up on any particular one. Also note that something that looks unpromising at first sight may end up being your favourite part of the course: geography has a knack of surprising you. Fans of the BBC comedy series *Detectorists*, written by and starring Mackenzie Crook, may be interested to hear that uniformbooks published a book in 2020 exploring the *Landscapes of Detectorists*. This saw academic geographers writing essays on the themes of each episode: some examples are gender, landscape change and interpretation, verticality, and the nature of the everyday that detectorists unearth through their hobby.

Think through possible research and career opportunities that may connect with particular modules. This may help you make a decision when faced with a choice between lots of modules. You may want to challenge yourself by picking a module that provides alternative views on the world. Interrogate the knowledge you are presented with and consider what sort of a geographer you want to be. Do you want to be someone who tackles the injustices that you encounter? Maybe you want to become someone who looks for solutions to problems? Or are you someone who would like to make new discoveries or take knowledge in new directions? Geography provides the scope for you to do all of these things. In a time of global crisis, geography will provide you with endless opportunity.

I'll come back to future careers in a later chapter, but it's worth considering these when picking modules, as it can inform your choices or affect the combinations of courses you should take. Maybe you'll want to ask your parents if they have any thoughts on appropriate careers. Would they make the same decisions again if they were back at university?

Mind the gap

One activity you may find it helpful to undertake for yourself is gap analysis, a technique developed by Joy Adams and Michael Solem from the Association for American Geographers. It involves choosing a possible occupation, identifying what you already know about it, and then researching it more fully. There is currently little specific career education provided in many schools – and even less on geographical careers. Take any opportunity you spot to find out more about the possibilities that geography provides. Solem suggests that you might want to discover more about

- the salary,
- the educational requirements,
- the opportunities for progression and
- the primary duties that are required.

These are then compared with your expectations to explore the gap between your perception of the job and the reality.[67] A small series of videos showing the work involved in several possible careers[68] can be viewed on the VirtualJobShadow website (https://virtualjob shadow.com/partners/esri).

The Royal Geographical Society (with IBG) provides detailed analysis of the cohorts that go on to take geography at different levels, including at university. The society also operates a long-standing 'Geography Ambassadors' scheme. This trains undergraduates to work with young people in schools to publicize the vital nature of geography and encourage the next generation of geographers to study the subject. Getting involved with the scheme could provide valuable experience to add to your CV – plus you get a nice t-shirt to wear! More importantly, you'll be asked to work in a school with a group of enthusiastic young geographers: you'll get to share your love for the subject as well as talk about your own experiences of studying and doing it.

Geography offers variety

Geography degree courses offer tremendous variety. There are hundreds of possible modules, and the examples below are just a few of the myriad modules that make up current university courses. I've tried to pick a selection that shows the great diversity of academic geography while sticking to topics that some of you may already be familiar with.

Bristol University offers a degree course connecting geography with innovation. It demonstrates the connection between the subject and intensely practical work. The course involves a great deal of independent work, in transdisciplinary teams. The following comes from the university's website:

> The innovators of the 21st century will bring together arts, science, engineering, humanities and enterprise to deliver innovative products, services and ways of living. They will be team players with a breadth of skills and qualities that enable them to work across specialisms and cultures.
>
> This course combines in-depth subject specialism in geography with interdisciplinary breadth, creative teamwork and entrepreneurial skills. Alongside your geography studies, you will apply your subject knowledge by translating ideas into plans for digital and creative enterprises, both social and commercial.

Leeds University offers an interesting array of second-year options, with students being able to focus their studies on a range of cities, including Helsinki, Montpellier and Belgrade.

Keep an eye out for 'psychogeography' modules as well. Phil Smith, Reader in Geography at the **University of Plymouth,** writes books in this area and connects geographical study with the performing arts. Site-specific art and poetry forms part of many arts festivals.

At **Keele University,** Peter Knight offers a second-year module called 'Inspirational Landscapes' that includes references to the music of Mahler, cowboy films and landscape painting. Students are encouraged to explore their own understanding of landscape in culture, which leads to a wide and interesting range of responses to Peter's teaching.

Staff at the Bartlett Centre for Advanced Spatial Analysis at **University College London** (www.ucl.ac.uk/bartlett/casa/) have carried out a range of interesting projects that use data and mapping to explore important areas of people's lives and use geospatial and open data creatively. Some of the results have been shared with the public. Their research is focused on 'the application of computer models, data visualisation techniques, innovative sensing technologies, mobile applications and urban theory linked to city systems'. The Colouring London website mentioned on page 78 is one of their projects, and they are also researching how an increasingly elderly population will be able to navigate and live in ever-growing urban centres. The geographies of play and of 3D mapping are two more of their areas of investigation.

Professor Ian Cook at the **University of Exeter** uses Lego to make political statements. He's interested in the ways that academics, filmmakers, artists, activists, musicians and journalists try to make tangible the lives of those who make and grow everyday commodities. He runs a spoof shopping website called Follow the

Things (www.followthethings.com) that tells the real stories of particular items (such as toys and fruit), and he has worked with Fashion Revolution to uncover the practices of the fashion industry, encouraging all consumers to ask who made their clothes. In conjunction with Finnish geographer Eeva Kemppainen, Ian developed ideas for 'subvertisement' workshops, where young people explore advertising imagery to develop their media literacy. The resulting guide is available in several languages.[69]

The **University of East Anglia** has a long-standing connection with climate change studies: its Climate Research Unit has operated in its own building since the 1970s, when Hubert Lamb led a department that is still well known worldwide today.

Phil Jones, who works at the **University of Birmingham**, explores urban regeneration and cycling. He's also interested in videogames and runs a 'Playful Methods Lab', which explores the use of virtual reality, among other things. His work challenges thinking on how academic geographers work.

The **University of Manchester** offers a module called 'Islands: human geographies of isolation' that explores the different ways that people use islands, from prisons to luxury holiday getaways.

❧

I have only scratched the surface here. Whatever your interest, there will be a geographical dimension to it, and it's likely that some academic will already have delved into it in great depth. Decide on the general area that you're interested in, whether that be human, physical or environmental in nature.

Geography and diversity

Geographers can also take pride in the fact that it's a very inclusive subject discipline, while acknowledging that, as with all subjects, further improvements can still be made to the subject's diversity, with all that entails. Diversity itself can be defined in many different ways. While geography certainly has something to offer everyone, not everyone, unfortunately, will always feel fully welcomed into all aspects of the geographical community. There is work to do here, too. Steve Brace from the RGS-IBG wrote an article in January 2020 in which he outlined some recent successes of geography in this area:

Since 2010 the subject's growth at GCSE (rising from 27% to 40% of the annual cohort) has occurred due to growing numbers of pupils taking the subject who were previously less likely to study it. These include Black, Asian and minority ethnic pupils, those who'd experienced disadvantage or lower prior attainment or were studying in comprehensive schools.[70]

A growing number of Chinese students are choosing to take A level geography in the UK. Many schools, particularly independent ones, have been successful in attracting overseas students, who can see the benefit of the subject for its facilitating nature when they move on to further study. They also get to improve their language skills as they study.

Gender is a regular feature of university modules, with cities being described as 'gendered' depending on the nature of their developments, the opportunities they offer their residents, and the barriers they place in the way of women. Many developments are designed around the able-bodied, so consider the exclusion of particular groups, which is invisible to many. The scope of the subject is such that sexuality and identity – including LGBTQ+ issues, gendered spaces and inclusion/exclusion – can all become a focus for further study and research. Regardless of age, gender, race or nationality, then, geography is very much a subject for you, although there is still work to do to decolonize the curriculum and ensure equal opportunities for all. Perhaps you can be an agent of change in this important work.

Organizations such as the Black Geographers group, created by geographer Francisca Rockey, are working to change the representation and engagement of minority groups within the subject community, and thinking about all that that involves. What similar opportunities exist within your own chosen institution?

Exam results for the 2020 cohort were subject to changes due to the cancellation of examinations and an eventual reliance on centre-assessed grades for A levels. This has led to an increase in the uptake for geography degrees, particularly by pupils from disadvantaged backgrounds.[71]

A final word is merited for one aspect of university life that is sometimes forgotten. In a world that has many challenges, studying for a geography degree offers you a rare chance to immerse yourself in the subject discipline for several years, to broaden your horizons and to fully engage with the subject's powerful knowledge.

Some things to think about before reading the next chapter

▶ Make an appointment with the student support officer who deals with careers advice in your institution and start taking steps towards securing a job, an apprenticeship or further study before your course ends.

▶ Take a look at the list of careers compiled by the RGS-IBG with a view to identifying some that you might consider for your future.[72]

CHAPTER 7

Why geography matters now more than ever

The study of geography is about more than just memorizing places on a map. It's about understanding the complexity of our world, appreciating the diversity of cultures that exists across continents. And in the end, it's about using all that knowledge to help bridge divides and bring people together.

— Barack Obama, US president 2009–17 (speaking in 2012)

IN THIS CHAPTER I WILL draw together some threads from previous chapters and explain why geography matters more now than it ever has. I'll remind you of some of the key reasons why studying geography is so important, not just for you personally, but for everyone in the world.

This is an important chapter, and I've been building up to it for a while. I hope you might already be able to suggest some reasons why geography matters more now than ever before, and this chapter will help to consolidate your thinking.

So why does geography matter now more than ever?

Put simply, geography helps us answer the great issues of our age. To use what might be called geological terminology, many scientists say we are now in the Anthropocene period. This is a period during which people are known to have had an impact on the surface of the earth and on other earth systems such as the atmosphere and cryosphere. Plastic has been found in every part of the earth, and microplastics are present in the food chain even in remote polar regions. We've changed the composition of the atmosphere by adding more greenhouse gases. Reports from numerous organizations warn us that we're changing natural systems for the worse. And geographers have been involved in the production of many of these reports: unsurprising given their curiosity about the world. The following quote comes from Danny Dorling and Carl Lee's *Geography* (2016, p. 7) and stresses the importance of curiosity:

> *Geographers have a tradition of being curious explorers both of places and ideas. Where does that highway go to? Who lives in this big house, and why? How did we arrive at where we are? When are we going to learn to live together? Can you really consume more and more, and does it really make you feel better? Is there something*

nagging away at you every day – a splinter in your mind – a thought
that somehow there must be a better way to live?

As the chapter title says, geography matters more now than ever.
The decades ahead of us are crucially important: global action is
needed, and we must start working on a global scale to confront
some of the 'wicked challenges' that face us. We need hopeful
geographies to emerge, with pragmatic solutions to these problems.
And alongside the global issues there are also local ones that require
some unravelling.

In late 2019 and early 2020, many residents of new homes in
the North of England found themselves affected by floodwater
following record-breaking rainfall. One would imagine that – given
the teaching of geographers over many decades – planners would
understand that land close to rivers might be prone to flooding. As
the late Philip Eden, geography graduate turned meteorologist and
weather presenter, said:

Floodplains are there for a reason – to permit periodic flooding when
rivers overtop their banks – and they are flat and easy to build on
because of the silt and mud that has been deposited there at regular
intervals over aeons of time.

And really, it should be apparent even to non-geographers that
floodplains are likely to flood, and it often is. Alan Titchmarsh, for
example, is a gardener with a 40-year TV presenting career behind
him. He previously presented the BBC series *British Isles: A Natural*
History. In December 2019, when the village of Fishlake near Der-
byshire was flooded, he said in an article in the *Yorkshire Post*:

Don't those words mean anything to anyone? Listen to me, I'll say
it again, 'flood plain'. Does that tell you anything at all? There are

reasons, for pity's sake, why places are called Foggy Bottom or Runnymede. It's because they are in a hollow where there's damp, or because there's a lot of water that collects on low-lying ground. It's as simple as that. So why ignore those ancient warnings?[73]

A little geographical knowledge can be powerful when making one of life's major decisions: where to live. It's not confined to these local decisions, though: it could actually unlock some of the main reasons for you, personally, to study geography.

Geographical knowledge is powerful knowledge

This idea of 'powerful knowledge' is an area that will already have informed your education so far without you necessarily realizing it, and it provides a particularly strong argument for geographical study.

Professor Michael Young has done more than most to popularize the idea of 'powerful knowledge'. He believes it to be the knowledge we need to go to school to obtain; in other words, it's the knowledge that we can't just obtain by Googling. This requires teachers to have both what is called 'powerful disciplinary knowledge' and the ability to share that knowledge with others. I hope you had an inspirational geography teacher – I've heard so many times from those who go into geographical study or careers that they were inspired to do so by their teachers (and the same goes for many other subjects too, of course).

Geography teacher Richard Bustin has researched powerful knowledge in geography. He says of the idea that it provides a sense of coherence:

Powerful knowledge accepts that new ways of seeing and writing the world are created by the geography academic community, which adds to and sometimes replaces discredited theories and ideas...

The task of 'making sense of the world' is the community endeavour of the 'discipline' with its own rules and procedures of selecting better knowledge.[74]

Your school studies have introduced you to the discipline and your teachers will have worked to create a curriculum that introduces you to key knowledge and concepts in a structured way.

Not only will geographical study give you access to the powerful knowledge that will help you understand the world, it will also help you make your own niche within it, in an uncertain future.

GeoCapabilities

In 2012, David Lambert, Sirpa Tani and Michael Solem, along with a number of partners, worked to develop an approach to geography education called GeoCapabilities.[75] It explains how schools should expand young people's capabilities, using the school curriculum (including geography) to enable them to think beyond themselves and their own everyday experiences. GeoCapabilities contributes to the 'substantive freedoms' available to young people: that is, the freedom to *think*, to *make good choices,* and to *make decisions* about how to live. The GeoCapabilities approach argues that an individual will develop greater potential to achieve these freedoms if they acquire geographical knowledge.

Being able to think geographically enhances the capabilities of young people in particular ways. This is regardless of whether or not they secure a geography-related job. We're all living geography every day, of course, but sometimes it's important to be reminded of this. In 2020, 'Why geography matters' was the theme of the Geographical Association's annual conference. Gill Miller's Presidential Lecture on this theme can be viewed on YouTube,[76] along with other sessions that took place.

Everyday geographies

Many of the world's major challenges occur as combinations of other factors. Think for a moment about some of those factors – such as inequality, conflict and migration – and how they might connect with your own daily lives. The technical term for this is the **quotidian**: the everyday experiences that we all have, many of which we don't even think about, as they are so familiar to us. As I've said, we're all living geography on a daily basis.

Open any newspaper today, in any country, and most of the stories in them will have geographical elements, or can be explained by 'thinking geographically'. Some stories will connect several different issues at varying scales, with one example being the bushfires in Australia, which dominated the news in December 2019. They linked global climate change with a regional atmospheric disturbance called the Indian Ocean Dipole, and they also involved more local actions relating to fire prevention, forest management and population density. Thinking geographically is a really powerful idea.

The Royal Scottish Geographical Society is of the opinion that 'everyone is a geographer'. It's an opinion I share, and my hope is that this book is helping you appreciate this for yourself. The section of the society's website that supports this campaign involves some useful resources that you could explore, and if you are anywhere near Perth (the city in Scotland, not the one in Australia) it's worth visiting their headquarters. The building is open to the public and hosts a range of fascinating exhibits from famous explorers who have been connected with the society over the last century: Sir Ernest Shackleton, Isobel Wylie Hutchison and many others are included.[77]

Young people's geographies

Powerful knowledge aside, it's worth remembering that adults aren't the only experts, and they're not the only geographers in the classroom either. It's entirely possible to develop geographical knowledge without having a teacher explain something to you: you can learn through discovery, or through personal experience and wider reading.

One Friday in August 2018, a 15-year-old girl sat by herself outside the Swedish parliament building in Stockholm with a hand-painted sign that read 'SKOLSTREJK FÖR KLIMATET' ('School strike for climate'). Her classmates wouldn't accompany her, so she went against the wishes of her teachers and her parents and skipped school and made her way to the city centre.

Over the next year, others gradually began to join her each Friday. Her message about the climate emergency began to spread widely, and she eventually became the figurehead for a global youth movement for change. Greta Thunberg (you knew it was her I was talking about, of course) has inspired millions of others to 'strike for climate' since that lonely afternoon in 2018. She has crossed the Atlantic to attend key climate conferences, speaking passionately to audiences at the United Nations and other global bodies about the way that those in power are ignoring the fact that their 'house is on fire'.

It's become clear over the years that Greta never wanted to be missing school, it was just that the cause was too important for her to ignore. So while she's undoubtedly missed out on a number of school geography lessons herself – she passed the hundredth week of her Friday strikes in July 2020 – she's been teaching the whole world a very important geography lesson in the meantime. In August 2020 she returned to school after a rather productive 'gap year'.

Greta isn't unique in having had an impact on other people in this way, either, and she wasn't the first young person to speak out about the climate crisis. Consider the impact of Malala Yousafzai, Autumn Peltier and Ayakha Melithafa, for example.

Dara McAnulty's book *Diary of a Young Naturalist* was awarded the prestigious Wainwright Prize for nature writing in 2020. He started writing the book, which describes the changing seasons, at the age of just 14, going on to receive the award when he was 16.

Geographical concepts are important

There are some key geographical concepts that you should become familiar with, preferably in advance of making important life decisions. Once they become a regular part of your daily thinking, you'll appreciate more keenly why geography matters. As we've mentioned, some geographers use the metaphor of the subject being a 'lens' through which they see the world. While there is no definitive list of key concepts that geography uses, those attempting to identify the most important themes generally include some or all of the following:

- place,
- space,
- scale,
- connectivity,
- complexity,
- interdependence,
- sustainability,
- environment,
- inequality,
- diversity and
- representation and perception.

An inclusive subject

Geography is a very inclusive discipline. Although the image of geographers that some may have is of bearded men in checked shirts, many female geographers have led the subject in new directions.

The Royal Scottish Geographical Society's Education Medal is named after Professor Joy Tivy, who worked as a biogeographer at the University of Glasgow and wrote textbooks that have been used by generations of students.

Professor Alice Garnett was one of the first female geography professors. She pioneered a number of new developments in the subject and was elected president of the Geographical Association in the 1960s.

Other recent presidents of the Geographical Association include Gill Miller in 2019 and Susan Pike in 2020. Both the current president of the RGS-IBG, Baroness Lynda Chalker, and the previous long-serving director Rita Gardner have held important positions in government and in other organizations.

Take some time to research gender equality in geography. See what you can find out about the following female geographers and campaigners, and note that they are just a few examples from many I could have chosen.

- Professor Patricia Daley, University of Oxford
- Professor Cindy Brewer, Penn State University
- Professor Katherine Brickell, Royal Holloway University
- Melati and Isabel Wijsen
- Wasfia Nazreen
- Marie Tharp
- Ella al-Shamahi
- Evgenia Arbugaeva

Untethered geographies

Social geographer Alastair Bonnett talks about the idea of 'untethered geographies': the revolution brought about by new technology that allows people to work anywhere. One interesting outcome of this phenomenon can be found on the Caribbean island of Bermuda. In 2020, Prime Minister Mia Mottley announced a new visa scheme called the 'Bardados Welcome Stamp'. This allows people to relocate there and work remotely for a year: an interesting option for those who are able to work in that way, and a scheme that was swiftly mirrored by other locations.

At a Prince's Teaching Institute event in June 2020, Bonnett explained that Google Street View is an example of the politics of visibility and inequality. The world's poorest are excluded from the mapping, as not all countries have been mapped with the company's cameras. The blank areas are usually outside of Europe: Figure 5 shows the countries that Google had mapped by 2017.

Which Countries Have Google Street View?
Countries and dependencies with full/partial Google Street View coverage (2017)*

* November 2017. Not including areas with views/photo spheres of selected businesses and tourist attractions

@StatistaCharts   Source: Google

statista

Figure 4 Countries covered by Google Street View by 2017. (*Source*: Statista. Available at https://bit.ly/2ZDtglX.)

Bonnett also reminds us of the importance of being connected to somewhere when he asks, 'Aren't the local places, the ones that are truly ours, the ones that matter most in a disorientated world?'

We've all been spending a lot of time at home during 2020, and this has made a real difference to people's awareness of their own place in the world. Geography is a subject that provides a wide range of opportunities for those who want to use it. Its value goes above and beyond the substantive contributions it makes to policy making at government level. It sits at the intersection of many other sciences and draws on them while adding its own particular viewpoint to the debate. It enables a critical understanding of the world. One way to think about the value of geography is to imagine what it would be like if geography *wasn't* part of the curriculum. As Alexander Murphy suggests:

> *Consider what is lost if geography is not part of the educational mix. Students may never be encouraged to develop even a basic understanding of how the world is organised environmentally, politically and culturally… Appreciating what geography has to offer requires moving beyond a shallow place-name-based conception of the subject that remains distressingly common among the general public.*[78]

The world's biggest companies understand the power of geography

You don't have to work for a geospatial company to be aware of the use of geographical location technology. Think about the location tools that allow customers to track their online purchase from the supplier's warehouse to their home. You're often able to see where the van carrying your parcel is and how many more deliveries it has to make before reaching you.

Delivery services use technology and algorithms to plot their vehicles' routes to reduce fuel costs and save time. In the US, for example, routes are designed so that UPS vehicles don't need to make turns across a flow of traffic, as this can cause delays at busy times. Online shopping increased in scale during the Covid-19 pandemic, as people turned to home-delivery options rather than shopping at bricks-and-mortar stores. This was particularly true for those who were shielding or self-isolating. Geographers understand that this will change the nature of our cities, post pandemic, in a myriad of ways that will unfold over the coming decade.

Michael Batty from the Bartlett Centre for Advanced Spatial Analysis predicts increased delivery-vehicle traffic, possibly augmented by drones, as well as more people working from home, perhaps reducing the need for people to travel to work on public transport. This would change the nature of offices and shift retail premises away from the centres of cities towards their edges, and it would result in growing demand for properties with outside space in rural areas.[79]

Uber provides transport options in various cities around the world. They're one of the world's biggest personal transport companies – and yet they don't actually own any vehicles. Their technology allows people to order an efficiently allocated vehicle, it provides customers with the reassurance that their journeys are tracked, it allows them to make payments and provide driver ratings, and it improves safety.

Similarly, Airbnb is one of the world's largest companies when it comes to offering people accommodation, but it doesn't own any property. Once again, it uses software to match property owners with prospective guests. Its popularity and success have led to a reduction in the availability of long-term rental opportunities for those who can't afford to buy a property, which has brought the company into conflict with the residents of cities such as Barcelona.

Meal-ordering apps such as Deliveroo and Just Eat were used by many families to order takeaway meals during lockdown. Many people working to deliver meals for these companies are part of what is called the 'gig economy': a new model for paying people for short-term work or freelance tasks rather than providing them with permanent employment. If you've used these services, there's no guarantee that your food came from an actual restaurant kitchen, as you might have thought when you ordered it. It might well have come from what is called a 'dark kitchen' instead. These are a relatively recent change to the culinary landscape: kitchens set up in temporary buildings, changing the 'usual' model of how we obtain restaurant meals.

Domino's sell pizzas but it is also essentially a logistics company. In order for it to run smoothly, both ingredients and staff have to reach the restaurant at the correct time. Considerably more effort goes into making this happen than goes into carefully positioning those pepperoni slices on your Half & Half.

You may not have heard of the company Buffaload, but if you shop at Co-op or Tesco, the items you've taken down from the shelves are likely to have been delivered by them. Their lorries have telematics fitted to track their positions and enable routing in the most efficient way. The company is also introducing various innovations in an attempt to become carbon neutral in their operations.

All of this helps build a picture that the value of geography goes beyond the confines of the specification that is taught in school or university. A geographically literate population would be less taken in by fake news, and it would appreciate the nuanced nature of much of the information it is presented with. It would understand some of the changes I have just outlined, and this would in turn help people make better decisions.

It's important to expand on people's geographical curiosity: to help them 'fill in the blanks', to appreciate difference and diversity, and to think about how issues might be viewed from elsewhere.

In a video he made for the students at my school during the 2020 lockdown, Joseph Kerski, esri's Education Manager, had a message for geography students:

We have confronted numerous crises around the world over the past six months, including those surrounding global human health, massive wildfires, and more. These crises show that understanding geography is more important now than ever before, and that by acquiring geographic content knowledge, perspectives, and skills in problem solving, using geotechnologies and other tools, and understanding the whys of where, you will be the change agents and decision makers that this world so critically needs.

Joe Smith, director of the RGS-IBG, reminds us again of geography's important position as part of the humanities:

The political tectonic plates will continue to clash and remodel our political landscapes. However, we should celebrate how study of the humanities can help us all to understand our own and other cultures, explore the connections of the wider world, and satisfy our innate curiosity about the world's people, places and environments.[80]

Think about how your geographical education equips you to question claims that are being made about the world. For instance, read Steve Brace's statement from the very start of this book again:

Young people are entering an environment of high tuition fees and a competitive job market, so they rightly want to know where geography can take them beyond school and university. For all students considering their next steps, remember that national statistics show undergraduate geographers are more likely than almost any other students to enjoy and complete their degree, and that geographers experience above average rates of graduate employment and earnings (for female geographers, that's up to 10% more than the average).

Could it be the case that in a very competitive job market, a geography degree – which, as we've seen, allows you to apply for many different jobs – is actually better than a degree that's more directly tailored towards entering a specific profession, especially if that profession might have dramatically slowed down its recruitment after the pandemic? Alternatively, ask yourself whether the fact that geographers experience above-average rates of graduate employment and earnings could be a result of advantages those geography students had before they took their degree, perhaps?

To finish, I want to return to the theme of exploration, as I believe that all of us are geographers and explorers at least to some extent. In his 1752 book *The Adventurer*, John Hawkesworth wrote:

> *Nature is now exhausted; all her wonders have been accumulated, every recess has been explored, deserts have been traversed, Alps climbed and the secrets of the deep disclosed.*

He was, of course, spectacularly wrong! Alastair Bonnett has called our age 'the era of exploration'. In October 2017, *National Geographic* explorer Daniel Raven Ellison wrote in a piece for *Geographical* that he felt that 'all geography teachers should teach their students to be explorers'.[81] I hope your teachers have done that for you.

I agree with Alastair that there's still plenty of the world where no one has stood. Exploration isn't necessarily about being the first to a location.

In his 2018 book *Geography: Why It Matters*, Alexander Murphy reminds us that geography is also addictive:

> *Exposure to geography can also serve to arouse interest in, and curiosity about, other peoples, places and landscapes... Knowing about something is a prerequisite to caring about it... Geography, when taught well, has the capacity to challenge physical and mental bubbles that constrain thinking and experience.*

One final important benefit of geographical knowledge might be to impress a potential partner. When the Duke and Duchess of Sussex visited Pakistan in 2019, the royal couple visited a glacier. Kate was said to be 'impressed by Prince William's geography'.

So please keep studying geography and exploring this amazing subject. We're not all 'prisoners of geography', but we should certainly remain in its thrall, always looking to rediscover the joy we experienced when, as children, we first saw the sea, or were let free of the pushchair's restraints, taking our first uncertain steps into a limitless world. Geography provides you with powerful knowledge. Use that knowledge to think in new and different ways about the world: think geographically. This may well give you the edge in our complex and ever-changing world.

I hope you're now ready to start making some important decisions about the next stage of your (geographical) life, and I hope the first such decision is that **geography is the right subject for you to pursue**. The current issues facing the world have placed geography firmly back on the agenda. I wish you the best of luck for the next stage of your education about the world.

Something to think about before reading the final chapter

▶ Now close the book and go for a walk. Perhaps a long slow walk, using the *Slow Walks* route map. Geography happens out there, beyond the four walls and the windows of your home. Go somewhere you've never been before. Never stop exploring.

CHAPTER 8

Finding out more about geography:
further resources and support

A passion for geography demonstrates a real curiosity about the workings of our wonderful planet and that is a great advantage in today's fast changing world.

— David Lambert, Emeritus Professor,
UCL Institute of Education

THIS CHAPTER PROVIDES A RANGE of suggested additional resources that you could read, view and listen to. They're intended to help you develop your geographical thinking as well as consolidate your decisions about where you might study geography. Plenty of ideas are offered for the long summer holidays that precede new courses, and I hope I can help you make a flying start in the next stage of your education. If you are moving into the world of work, this chapter should help you to continue to understand the dynamic world we live in and will encourage you to remain a lifelong student of geography. For those planning to undertake further study, there's a lot to explore here, and there's plenty for parents to read and watch too if they have caught the geography bug.

Geography is a visual subject, and those who study it have a huge range of resources at their disposal: from maps to imagery, satellite data, infographics, music and sound recordings. Many different media have been used to communicate the importance of geography. Many different 'geographies' exist. And there are many different stories. The various lists that follow introduce you to a few personal favourites that I've selected to help you take your thoughts beyond the pages of this book.

Useful books with a geographical theme

There are many books that contain geographical themes and that may provide you with some inspiration. Every geographer would benefit from having the selection that follows on their shelves.

Factfulness *by Hans Rosling (Sceptre, 2018)*
The late Professor Hans Rosling wrote *Factfulness* with his son Ola and daughter-in-law Anna to explain the titular concept, which they describe as 'the stress-reducing habit of only carrying opinions for which you have strong supporting facts'. Geographers are

factful people. The Gapminder website, founded by Rosling, is also excellent to support the book's key ideas: a series of 'rules of thumb' you should apply when presented with information about the world. The key idea here is that you should be critical of the information presented to you in the media.

Adventures in the Anthropocene *by Gaia Vince* *(Chatto and Windus, 2014)*

Many scientists believe we're now in a new geological era: one in which humans have changed the landscape, leaving an indelible mark on it, whether as plastics in the sediments of the ocean floors or as gases in the atmosphere. Vince travelled the planet uncovering evidence and identifying particular places that humans were changing for the worse, and occasionally for the better. The book charts a fascinating journey of discovery. Vince's more recent book *Transcendence* explores what it means to be human, looking at the power of language and stories and at how we evolved as a species.

Sapiens: A Brief History of Humankind *by Yuval Noah Harari* *(Harper Collins, 2015)*

Harari's best-selling book explores the history of our species and its development, and the resulting impact that we've had on the landscape. His book *21 Lessons for the 21st Century*, which explores similar themes, is also worth checking out. Both books are well written and show the scope of the subject, although neither is explicitly a book about geography.

No One Is Too Small to Make a Difference *by Greta Thunberg* *(Penguin, 2019)*

Here you'll find a collection of speeches made by Thunberg during her rise to prominence in 2019 as her climate strike caught the attention of the world. You'll find good examples of the simple messages that Thunberg used to try to make the older generation understand that their house was 'on fire', and that while they needn't be listening to her, necessarily, they should be paying attention to the facts about climate change. She's since developed her campaign, which will continue to push everybody to change their behaviour. The pandemic reminds us that to solve 'wicked problems' such as the climate emergency, a concerted global effort will be required.

Underland *by Robert MacFarlane (Hamish Hamilton, 2019)*

I recommend all of MacFarlane's books, which are learned and extremely well written. Although each has its own particular focus – mountains, walking, natural vocabulary, paths and tracks or, in this case, the subterranean – they are all useful in developing a holistic view of our relationship with nature and the vocabulary that can be used to describe it (earth writing). *Underland* is described by the author (an English specialist at Cambridge University) as a 'deep time journey'. It looks at the hidden geographies beneath our feet, including explorations of cave systems and mines alongside the 'wood wide web' that involves connections and communication between trees.

Future Fossils *by David Farrier (2020)*

As mentioned earlier, we live in a time when humans are leaving their footprints on the world. They can be found in the plastic laid down in ocean sediments, the gases that we have placed in the atmosphere, the radioactive material that is being locked away in repositories for thousands of years, and the changes we have made to the planet's surface.

As with some of the other authors in the list, Farrier is not a geographer, but his thinking is geographical in nature and also futures oriented.

Geography *by Danny Dorling and Carl Lee (Profile Books, 2016)*
This short manifesto for the importance of geography explores why the subject matters and identifies a number of key concepts that underpin a great deal of its power, explaining why they are important along the way. The book's sections on globalization, inequality and sustainability explain the global reach of geography. It's clearly and accessibly written, and nicely illustrated with Ben Hennig's cartograms.

Geography: Why It Matters *by Alexander B. Murphy (Polity Press, 2018)*
This is another quick read – and one with a bright pink cover. Murphy moves quickly through his explanation of why geography matters, drawing on a range of sources and providing useful summaries of key geographical thinking. The arguments are backed up with numerous examples from around the world.

Off the Map *and* Beyond the Map *by Alastair Bonnett (both Aurum Press, 2014 and 2018)*
Bonnett describes his visits to a range of locations that he's experienced in both traditional and more 'psychogeographical' ways. His books look at unusual places that have been created – or are still being created – by recent geographical processes and trends. Some of these places are real, others are imaginary.

The World: A Brief Introduction *by Richard Haass (Penguin, 2020)*
Haass is the president of the non-partisan Council on Foreign Relations, and his book is an excellent introduction to geopolitics for A level and undergraduate geographers. The book develops some of the themes in Tim Marshall's *Prisoners of Geography*, considering how different regions may change in the coming decades and explaining some of the context for why they're in their present state.

Social media and online resources

Consider using the following list to set up your own curated account for your geographical thinking. This can be shared with others, who you could also collaborate with.

Instagram

Instagram is used daily by more than a billion users around the world, sharing images of the quotidian or the everyday (www.instagram.com). There are many feeds worth following (that don't just involve pictures of food), with a wide range of travellers and explorers sharing photos of the places they've visited. You may also find it helpful to start your own account, sharing images you've taken yourself. You'll find a community of geographers, travel writers, outdoor bloggers and others who will provide you with inspiration while also being an audience for the things you share.

Facebook

Facebook allows users to create a page or set up a group based around a specific interest (www.facebook.com). There are plenty of these that have been created to share stories of geographical interest. You can set up your own page quite easily or join those of other geographers with a mutual interest. You may also find a page sharing stories about your local area, or about particular aspects of geographical study. Most school-level geography qualifications will have a suitable page, as will almost all university courses.

Hashtags

Hashtags provide another way of searching for and exploring geographical themes. Try adding #geography, or something more specific, to the search terms you're researching. With such a vast number of social media posts being created each day, you'll inevitably find something of use – although it will undoubtedly sometimes involve seeking out a rare diamond in the rough. One search term to follow is **#GenGeo** (for 'Generation Geography'), which shares the work of the current generation of geographers across various related disciplines. Also check out **#teamgeography** and **#geographyteacher** – or perhaps start your own.

Twitter

You may not have previously thought of Twitter as being of educational value, but many people who use it like the immediacy of the contact with others that it offers (www.twitter.com). It's also potentially a way for you to ask university academics questions, as many make use of it as a way of communicating their research interests and their work. I'll suggest a few UK university geography department accounts to follow later. These can be added to your account, or you can make a 'List' of relevant accounts as a curated news feed connected with your studies.

Pinterest

Pinterest is used to curate galleries of images sourced from web pages (www.pinterest.com). Images that you come across while browsing can be saved to your account using an extension to your web browser. They can be added to themed galleries called 'Boards' for projects that you're working on, or for essay themes. This can be useful for any research you're doing, and it can provide inspiration later, reminding you of the visual nature of geography.

Flipboard

Flipboard buttons installed in your web browser allow you to quickly capture links to full articles and create online magazines on a number of topics (www.flipboard.com). Both Flipboard and Pinterest allow you to share and collaborate. They're free to use and they're accessible via smartphone apps: great for capturing images and stories on the move. You may find yourself with more articles than you could ever possibly read – there's a lot of geography out there!

Geographical

The *Geographical* journal feature called 'I'm a geographer' appears in each month's issue of the magazine. It will help you learn about the variety of people who consider themselves to be geographers, not all of whom could be described as obvious candidates. The journal's website offers you a chance to read previous entries: https://geo graphical.co.uk/people/i-m-a-geographer.

The Conversation

Many of the articles in *The Conversation* are linked to geographical themes: see https://theconversation.com/uk. The articles are concise, can be shared under Creative Commons, and could help broaden your understanding of some key ideas that may crop up in interviews or as part of future courses.

Blogging

Have you ever thought about starting a blog? This needn't involve you spending any money: platforms such as Blogger, Wordpress, Wix and others allow you to start sharing your ideas, writing or images in minutes, for free. Good blogs can find a global audience, and if you add analytics to measure your global reach, you could use it as an electronic portfolio for future job applications.

Flickr

Flickr is another photo-sharing website that offers free storage for your images (www.flickr.com). They can be arranged into albums and shared with others. Remember to use your camera regularly (most likely the one on your phone), and share the resulting images using a Creative Commons license.

Geography websites for subject knowledge

Time for Geography

Time for Geography is the work of a group of geographers, including teachers, academics and filmmakers, who work together to create videos on geographical themes for students and their teachers: see https://timeforgeography.co.uk/. If you're registered with their website, you can save videos to watch later. They've produced a growing number of videos that introduce the work of particular university departments: these may provide you with inspiration if you find a course that looks of particular interest. They've won awards for the quality of their videos and for their work with a growing number of geographical organizations.

TED Talks

There are more than 3,000 TED Talks available to view online. Most come with annotated scripts and subtitles in many languages, and they're usually available in a range of formats. Search through the talks at www.ted.com/talks or use the site's recommendation tool. Here's a list of talks with a particular geography focus that I'd recommend.

- Ella-al-Shamahi: 'The fascinating (and dangerous) places scientists aren't exploring'
- Parag Khanna: 'How megacities are changing the map of the world'
- Michael Bierut: 'The genius of the London Tube map'
- Balsher Singh Sidhu: 'Are we running out of clean water?'
- Chimamanda Ngozi Adichie: 'The danger of a single story'

- Shashi Tharoor: 'Why nations should pursue soft power'
- Tristram Stuart: 'The global food waste scandal'
- Danny Dorling: 'Maps that show us who we are (not just where we are)'
- Carolyn Steel: 'How food shapes our cities'
- Hans Rosling: 'The best stats you've ever seen' and 'The magic washing machine'
- Luisa Neubauer: 'Why you should be a climate activist'
- George Monbiot: 'For more wonder, rewild the world'

An excellent addition to the geography scene came along in April 2020, from an unlikely source. Oliver Jeffers, the celebrated children's author, wrote a book called *Here We Are: Notes for Living on Planet Earth*. His 'Ode to living on earth' includes some interesting geographical perspectives on our home planet: I urge you to check it out.[82]

Streaming services

There are many programmes related to geography, or with geographical themes, on streaming services such as Netflix, Amazon Prime or BBC iPlayer. Some of the available shows are collated by teachers on Twitter every week, providing a schedule of suitable viewing. A search using the #geogglebox hashtag will lead you to lists of geography programmes for the week ahead that have been put together by various school geography departments. You'll always find something of interest to the curious geographer.

World Economic Forum

The World Economic Forum's website is regularly updated with analysis and further suggestions for discussion: see www.weforum .org. They also publish numerous articles and have a useful YouTube channel: www.youtube.com/user/WorldEconomicForum.

Gapminder

We mentioned the Gapminder Foundation and the important work of Hans Rosling in counteracting global ignorance earlier. The associated website (www.gapminder.org) hosts the very important Gapminder Tool, which allows you to explore a wide range of global data using 'bubbles' representing each country. Key indicators can be selected, and a time slider tool then allows you to check the relative progress made on each indicator through the last 200 years.

Dollar Street

A website related to the item above is Dollar Street, created by Anna Rosling-Ronnlund: see www.gapminder.org/dollar-street. The site contains more than 30,000 images taken in hundreds of homes in countries around the world. The houses are arranged on a nominal street, with the poorest at one end and the richest at the other. It's a fascinating tool for exploring what life is like for people on different income levels and seeing what they're able to afford.

Worldmapper

Worldmapper, which we've mentioned previously, hosts hundreds of maps and cartograms on its website at https://worldmapper.org/: they'll broaden your ideas by focusing on a specific variable and illustrating how it varies around the world by changing the relative size of each country.

Our World in Data

Our World in Data is the work of Max Roser and colleagues. The project's website (https://ourworldindata.org/) and Twitter feed (@OurWorldinData) share maps and data to help explain the current state of the world.

ArcGIS Online and StoryMaps

If you're a student, you can obtain a free account enabling you to use esri's professional mapping tools to explore the world. StoryMaps are great for connecting text, images and maps to create a narrative. They're helpful for student coursework as well, and plenty of guides are available to help you learn how to use them.

Young Geographer of the Year

Could you become a Young Geographer of the Year? There are several competitions you can enter that will give you an opportunity to practise your geography skills, but the RGS-IGB's annual competition to find the Young Geographer of the Year is perhaps the best one for an aspiring young geographer. Previous winners can be viewed on the associated website,[83] along with the recent topics (which are announced annually). Recent ones have included 'The Arctic' and 'Geography beyond the window'. The RGS-IBG also offers an award for the best 'non-examined assessment' (at A level) of the year.[84]

Novels, movies and TV series of interest to geographers

If you'd rather read a fiction book than a factual one, here are three books that have a geographical slant and are worth your time.

America City by Peter Beckett (Corvus, 2017)

America City is a science fiction novel set in an alternative future. It finds the American president trying to secure new land in Canada as the earth's climate warms, trying to prevent Mexican migrants from entering the country, and trying to stave off the chaos occurring in other parts of the world. It's a well-written and pacey piece of fiction that has connections to many topical themes.

The Wall *by John Lanchester (Faber & Faber, 2019)*
The Wall explores an alternate future in which the citizens of an unnamed country are forced by law to spend time manning a wall that's been built around the country's coast to help keep out immigrants using military force. I won't spoil the book for you by revealing the twist about which country it is or by telling you how it turns out for the protagonist.

Dark Matter *by Michelle Paver (Orion, 2010)*
Dark Matter is described as a 'ghost story' and is set in the far north of Svalbard, where a group of scientists spend the winter months in a remote cabin in the wilderness. As the winter darkness descends, strange things begin to happen in their isolation. It provides an excellent sense of place for this Arctic location, and the changing weather conditions as the winter arrives are wonderfully described.

Disaster movies

A fun way of testing out your geographical thinking is to watch a disaster movie and critique the accuracy of the science and the events that are being shown. With a little inside knowledge, the failings of many of these films become very obvious, very quickly. There are many candidates for you to try this with, but you could start with *The Day After Tomorrow*, *Twister*, *Dante's Peak*, *The Core*, *Geostorm* and *San Andreas*. As a rule of thumb, if Dwayne 'The Rock' Johnson is in a film, you may have to suspend your disbelief! There's a list on the Letterboxd website of 'Bad Geology Films' that you might want to check out,[85] or simply tune into the Horror Channel on Freeview, which shows films like this most days.

David Attenborough

For many people, Sir David Attenborough is the person who explains the natural world best, and in the clearest way, especially when he's accompanied by the spectacular filming of the BBC Natural History unit. Recent relevant series include *Blue Planet II* and *Planet Earth II*. Stories about ocean plastics and climate change have become increasingly important in recent series, and the work of the film makers has become increasingly focused on these kinds of issue. When the lockdown started, the BBC even drafted Attenborough in as a geography teacher. The historians ended up with Danny Dyer. One–nil to us, I'd say.[86]

The Impossible *(Warner Bros. Pictures, 2012)*

The Impossible is based on a true story of a family caught up in the Southeast Asian tsunami of Boxing Day 2004. The film has some harrowing scenes, but the tsunami itself is very well rendered, as the water sweeps through the coastal resort and carries members of the family far inland. See what you can find out about the true-life story of Tilly Smith, who proved that good geography teaching saves lives.

Hotel Rwanda *(United Artists/Lions Gate Films, 2004)*

Hotel Rwanda tells the story of the 1994 Rwandan genocide. After decades of growing tension between two tribal groups, the Hutu and the Tutsi, conflict erupted, and a tenth of the country's population was killed during a period of just 100 days. You can read more about Rwanda's recovery, and the unusual vehicle that helped some people overcome their issues (the bicycle), in Tim Lewis's book *Land of Second Chances* (VeloPress, 2013).

Ghosts of the Tsunami *(Macmillan, 2017)*

Ghosts of the Tsunami is a sophisticated and challenging book. It describes the impact of the 2011 tsunami on Japanese society along with the actual physical impact of the wave itself. It contains some harrowing scenes alongside descriptions of landscapes and communities torn apart and having to rebuild, and reminders of previous disasters and how they have shaped the character of Japanese people and society. You will learn that the impact of a tsunami is not just water.

Parts Unknown *(Zero Point Zero for CNN, 2013–18)*
Anthony Bourdain was a chef who made his name writing about the realities of life in his profession. In later life, he made a series of programmes about his travels to unusual destinations to (often) eat unusual foods. The shows provide a view into the real lives of many peoples and their culture. The series of travelogues, called *Parts Unknown*, shines new light on a number of countries around the world, many of which have had challenging recent history.

When you next sit down to watch a movie, I challenge you to find one that doesn't have a variety of geographical themes, either explicitly or otherwise.

Geographical lectures and podcasts to listen to

There's a world of audio out there that might help you unlock ideas about where geography could take you. You'll find things that are perfect to listen to when travelling on a train or as a car passenger, and others that you might want to listen to in bed. While I'd like to think they won't send you to sleep, they might be perfect for those needing a late-night voice for company.

Search the podcasts section of your preferred podcast service using appropriate geographical search terms. There are many series and individual podcasts that could be of interest, some of which are created by university academics or schools. This is a very democratic medium. With a simple app like Podbean, you could even make your own podcast to share your personal geographical discoveries. Get started by checking out the podcast series 'Living Adventurously' from the author and explorer Alastair Humphreys, for example.[87]

Radio 4's *50 Things that Made the Modern Economy*, by Tim Harford, has entries on the GPS handset, plastic and other items with geographical connections.[88] The episodes are short and snappy. Harford recently completed a follow-up, *The Next Fifty Things that Made the Modern Economy*, which continues on the same theme.

The RGS-IBG has a very useful 'Ask the geographer' series of podcasts that can be listened to online via Soundcloud. A large number of episodes are available on a range of themes: maybe you could start with Tim Marshall on borders, or the use of graphic novels to help explore disasters and the carbon cycle?[89]

The Geographical Association has produced its own series of podcasts called GeogPods. Some of them have more of a teaching focus, but the series is growing in scale and range over time. They can be found on the association's website[90] or downloaded with an app such as Podbean, to be listened to on your smartphone, on the move.

The BBC adds new episodes to its *Costing the Earth* radio programme each week. Episodes often have geographical themes, including plastic pollution, climate change and deforestation.[91] In April 2020, the programme explored decisions by arists such as Coldplay, The 1975 and Billie Eilish to move towards greener touring and to reduce their carbon footprint. This is an area that will continue to develop, given the virtual nature of many musicians' recent contributions.

The Inquiry, a weekly BBC World Service programme, 'gets beyond the headlines to explore the trends, forces and ideas shaping the world'. Podcasts of each episode can be downloaded from the BBC website.[92]

The Documentary is another BBC World Service programme that explores geographical themes linked to development and global issues. It's good at providing an alternative perspective on global themes.[93]

Gresham College has a series of lectures that you can watch online (www.gresham.ac.uk/watch/). These are delivered by subject experts and are quite lengthy. Sample lectures include Professor Jacqueline McGlade on whether gender equality can help solve climate change.

Many events were released in a digital format, or became virtual, as a result of the coronavirus pandemic. In some cases, the results can still be seen online: for instance, you can currently watch a series of talks from the Hay Festival Digital, many of which have geographical themes. A subscription area offers the full range of topics, but free talks are made available for a limited time as tasters for this offering (www.hayfestival.com/hayplayer/).

One idea is to install the BBC Sounds app on your smartphone, which will allow you to access radio shows and podcasts more easily. It's free to download from all app stores. You'll also find it useful if you become a student, although you will need to have a TV license if you want to access programming live: another expense adding to the student experience when you leave home.

Libby is a free app for your smartphone that connects with your library account. It allows you to take out audiobooks or ebooks free of charge, without the need to visit the library in person. If you aren't already a member of your local library, you should be!

There are several Netflix shows that you'll find relevant. A series made by Vox called *Explained* is made up of short episodes on themes such as 'The future of meat' and 'The world's water crisis'. The series *History 101* should actually have been called *Geography 101*. It has short episodes on topics such as plastics, the rise of China, and fast food. Another good series is Zac Efron's *Down to Earth*, which features a number of travelogues.

Finally, a series of videos that is well worth checking out are those made by 'The Curious Geographer': see https://bit.ly/33n8bgy. It features discussions on geographical subject areas, book reviews and revision sessions, and discussions with academics from a number of university departments.

Awarding bodies

The organizations listed below are the awarding bodies for UK schools that offer geography qualifications at GCSE and A level (as well as other international qualifications). They each offer subject-specific areas.

- AQA: www.aqa.org.uk/subjects/geography
- Edexcel: https://qualifications.pearson.com/content/demo/en/subjects/geography.html
- OCR: https://ocr.org.uk/subjects/geography/
- WJEC: www.wjec.co.uk/qualifications/
- Eduqas: www.eduqas.co.uk/qualifications/
- International Baccalaureate: www.ibo.org/
- Cambridge Assessment: www.cambridgeassessment.org.uk/
- Cambridge Technicals: www.ocr.org.uk/qualifications/cambridge-technicals/
- BTEC: https://qualifications.pearson.com/en/about-us/qualification-brands/btec.html
- EPQs are offered by several awarding bodies

Each organization has subject-specific areas on its website, and they all provide updates on changes to the examination format as a result of Covid-19 as well as detailing the longer-term impact of the pandemic on public examinations.

References and further reading

Who to follow?

The Geographical Association is the subject association for geographers: see www.geography.org.uk. It was founded in 1893 by a group of university academics and teachers who wanted to share resources. Its mission is to further geographical knowledge and understanding through education. Perhaps the most relevant aspect of its work is its network of regional branches, which are dotted all over the country and offer lectures from examiners, teachers and academics. Find out if there is a branch in your local area and check the programmes online.[94]

You can follow the association on Twitter using @The_GA, and you'll find separate accounts for its committees and interest groups, and for some of its branches.

Geography Education Online (GEO) is an innovative new website (https://geographyeducationonline.org) created by the Geographical Association for students studying for a geography GCSE or A level, or equivalent qualifications. GEO is designed to offer young geographers a new and engaging way to extend their geographical thinking and support examination success in 2021 and beyond.

The RGS-IBG (www.rgs.org) is the learned/royal society for geographers; it has had a particular focus on exploration since 1830. The society holds regular free exhibitions of photography, artwork and artefacts from its collections in the foyer of its headquarters, which is close to the Royal Albert Hall and the Science Museum. These are usually free of charge. Members can view filmed versions of the society's Monday night lectures by logging in; note that there is also a 'Young Geographer' membership, which provides access to these lectures and other special events. The lectures are delivered by experts in their field, many of whom are household names (e.g. Michael Palin). They would be a particularly valuable resource to anyone who finds themself studying at a university in London (of which there are several).

You can follow the society on Twitter using @RGS_IBG or @RGS_IBGschools. It's also worth following @GeogDirections for its blog.

University geography Twitter accounts worth following
Many UK universities have active social media feeds for geography courses. The following list gives some examples, but it's not intended to be comprehensive.

- Birkbeck, University of London: @bbkgeography
- University of Birmingham: @GeogBham
- University of Bristol: @GeogBristol
- University of Cambridge: @CamUniGeography • @cambtweetgeog
- Cardiff University: @CUGeogPlan
- University of Chester: @gidchester
- Durham University: @GeogDurham
- University of Edinburgh: @ GeosciencesEd
- Exeter University: @ExeterGeography • @SaffronJONeill • @followthethings
- University of Glasgow: @UofGGES
- Imperial College London: @ImperialRSM
- King's College London: @kclgeography
- Leeds University: @SoGLeeds • @gisatleeds • @alexschafran
- Leicester University: @LeicesterGeog • @CCGLeicester
- Liverpool University: @livunigeog • @alexsingleton • @geodatascience
- London School of Economics: @LSEGeography

- Manchester University: @GeographyUoM
- Maynooth University: @Maynoothgeog
- University of Northampton: @GeogNorthampton
- Nottingham University: @UoNGeography
- University of Oxford: @OxHumanities • @oxford_ageing • @dannydorling
- Queen Mary University of London: @QMULgeography
- Queen's University Belfast: @QUBGeography
- Royal Holloway, University of London: @RHULgeography
- Sheffield University: @undertheraedar
- Staffordshire University: @StaffsGeography
- St Mary's University: @StmarysGEOG
- University of Sussex: @SussexGeog
- Trinity College Dublin: @TCD_geography
- University College London: @UCLgeography
- University of York: @YorkEnvironment

Mainstream media and journals with a geographical theme

It would be worth considering a subscription to a relevant journal or magazine. Perhaps you could request it as a Christmas or birthday present that would continue providing you with a monthly gift the whole year round. Many journals are available digitally to be read on your phone or tablet.

You may be invited by your teachers to subscribe to journals such as *Worldwise* or *Geography Review*, both of which are published by Philip Allan. Your school will be able to access these at a reduced price, and they both contain useful articles as well as advice on how to answer exam questions. Many journals can often be found in school libraries, or in the local public library; they may have a wider range than the local shops and will have the advantage of being free to access.

Some journals with strong geographical content include the following:

▶ *The Economist* contains a wealth of articles and useful illustrations and infographics. It is available in print or as an e-journal: www.economist.com.

▶ *New Scientist* is available in print or as an e-journal: www. newscientist.com. It shares the latest scientific stories, many of which are related to the environment.

▶ *Geographical* – the magazine of the RGS-IBG – is published monthly and is available in print or e-journal formats (https://geographical.co.uk). It is one of the more accessible journals and includes a monthly column in which the best-selling author Tim Marshall explores global issues.

▶ *Weapons of Reason* is a series of eight themed journals published by Human after All (https://weaponsofreason. com). Back copies and digital versions of some issues are available, with themes including the Arctic, artificial intelligence, food, inequality and urban life.

▶ *National Geographic*, with its distinctive yellow-framed design, has been published for more than a century (www. nationalgeographic.com). It contains a great range of articles and has some good online-only content, too: check out 'What Is Geography' on their website.[95] Hundreds of hours of National Geographic TV programmes are also available on the Disney+ subscription streaming service, which launched during 2020.

▶ *Monocle* is a pricey but well-designed monthly journal that gives a valuable new perspective on our lives, particularly on urban living, alongside other content. Its website is also well worth exploring (https://monocle.com/magazine).

▶ *The Spectator* is another magazine that comments on political events with a geographical theme (www.spectator. co.uk). It also has an interesting podcast series.

▶ *Vox* is a website based in the US, but it has a wide-ranging scope (www.vox.com). It offers articles and explainer pieces and hosts useful videos on its YouTube channel. As with all the sources listed here, geographers will be aware that a particular political leaning is something to look out for.

▶ *Ernest* is a magazine with a focus on the outdoors. It is published infrequently, but it has a nice design and contains interesting pieces of writing that could act as a model for your own creative writing on the environment, so look out for it. Another similar travel-based journal is the magazine *Suitcase*.

Some places to visit

Remember that all universities offer open days for potential students, and all universities are interested in attracting you to study with them. Prepare some questions – many of which could be 'geographical' in themselves – to help you find out more about potential universities if you're going to commit to spending three years in one place. Most universities produce promotional videos or provide virtual reality tours of their accommodation and lecture theatres. These can help you get a feel for a place without having to pay for a train ticket.

CHAPTER 8 – FINDING OUT MORE ABOUT GEOGRAPHY

There are also several useful regional events and conferences that include sessions with geographical themes: for example, the city of Cambridge has both a Festival of Ideas and an annual Science Festival, as does York.

Visit relevant museums, e.g. the Museum of London, the Natural History Museum, the V&A, the Science Museum, many of which are free. Your own local town museum is likely to have relevant displays as well. Keep an eye out for art projects that are connected with the landscape and environment, which you might want to participate in during holidays or on your own time. These might help you develop your own skills, and they'll also potentially be valuable for your CV in the longer term.

Visit your local National Park and investigate the landscape and how visitors are managed. How do the visitors impact the landscape and other activities within the park? Can you see signs of this? Are there conflicts between residents and visitors?

University applications

- Higher Education: https://hecsu.ac.uk/
- Discover Uni: https://discoveruni.gov.uk/
- UCAS: www.ucas.com/
- Informed Choices (Russell Group Universities guidance): www. informedchoices.ac.uk/

Careers guidance

High Fliers Research produces an annual report about the graduate market that includes some insight into the value of particular subjects when it comes to graduate recruitment. Its 2020 report can be downloaded as a PDF from www.highfliers.co.uk. And one final useful resource is the government website dedicated to apprenticeships advice (www.apprenticeships.gov.uk).

ENDNOTES

1 Steve Brace, 2019, 'GCSE results: "A tectonic shift in geography entries"', *Tes*, 22 August (https://bit.ly/3iszUmg).

2 Kathryn Snowdon, 2019, 'Brexit and climate change said to be behind surge in politics A-level', *NCFE*, 15 August (https://bit.ly/35B87wy).

3 David Wolman, 2020, 'Amid a pandemic, geography returns with a vengeance', *Wired*, 14 April (https://bit.ly/2RpsAvP).

4 See the website of the Department of Geography and Environmental Engineering, Johns Hopkins University, Whiting School of Engineering (https://bit.ly/35zkSrl).

5 Mark Jones (ed.), 2017, *The Handbook of Secondary Geography* (Geographical Association).

6 Sophie Donovan, 2019, '*Prisoners of Geography*: an interview with Tim Marshall', *Geographical*, 30 October (https://bit.ly/3kiJqZY).

7 Source: www.britannica.com/science/geography.

8 Noel Castree, 2005, *Nature* (Routledge).

9 Alexander B. Murphy, 2018, *Why Geography Matters* (Polity Press).

10 Danny Dorling and Carl Lee, 2016, *Geography*, Ideas in Profile (Profile Books).

11 Graham Butt, 2011, *Geography, Education and the Future* (Continuum).

12 S. James, 1990, 'Is there a "place" for children in geography?', *Area* **22**(3), 278–283.

13 Alan Gussow, 1997, *A Sense of Place: The Artist and the American Land* (Kogan Page).

14 Doreen Massey, 1995, *For Space* (SAGE Publications).

15 Source: www.geography.org.uk/students-views-of-geography.

16 Archibald MacLeish, 1968, 'A reflection: riders on earth together, brothers in eternal cold', *New York Times*, 25 December (https://nyti.ms/3cYncZH).

17 Chimamanda Ngozi Adichie, 2009, 'The danger of a single story', *TED Talk*, July (https://bit.ly/2E47ZKE).

18 Vicki Phillips, 2020, 'Why geography should be part of every 21st century education', *Forbes*, 31 August (https://bit.ly/3kdWEqG).

19 'Private school pupils earn more than state school pupils by age 25, IOE research reveals', Institute of Education, University College London, 8 November 2019 (https://bit.ly/33mcadi).

20 'Gender inequality and women in geography', Royal Geographical Society (https://bit.ly/33vS2W2).

21 Success at School: https://successatschool.org. The geography-specific page can be found at https://bit.ly/3bYaV8c.

22 'Choose geography at school', Royal Geographical Society (https://bit.ly/2Fr3aeY).

23 John Muir Award: www.johnmuirtrust.org/john-muir-award.

24 Eco-Schools: www.eco-schools.org.uk/.

25 '*The Guardian* view on geography: it's the must-have A-level', *Guardian* editorial, 13 August 2015 (https://bit.ly/2RCG8V7).

26 See note 20.

27 'Geography teaches the next generation about climate change', Royal Geographical Society (https://bit.ly/2DZo4kL).

28 Source: https://luminate.prospects.ac.uk.

29 Source: https://luminate.prospects.ac.uk/what-do-graduates-do.

30 London Economics, 2020, 'Qualified for the Future: Quantifying demand for arts, humanities and social science skills – May 2020', Education and Labour Markets, 7 May (https://bit.ly/3mgbejr).

31 Jon Pickstone and Rita Gardner, 2017, 'GSE to recruit new head of geography', *Government Science and Engineering*, 25 July (https://bit.ly/2FvnnjR).

32 From the esri User Conference Plenary, July 2020.

33 Source: https://teach-with-gis-learngis.hub.arcgis.com/.

34 The Royal Society, 'Dynamics of data science skills: how can all sectors benefit from data science talent?', Report, 9 May (https://bit.ly/2Fquzxu).

35 Mark Gurman, 2020, 'Apple, Google bring Covid-19 contact-tracing to 3 billion people', *Bloomberg*, 10 April (https://bloom.bg/3hAURKM).

36 Joe Smith, David Cannadine and Norman Gowar, 2019, 'Securing a future for humanities: the clue is in the name', Letter to *The Guardian*, 18 March (https://bit.ly/2ZD8X7Y).

37 Murphy (2018): see note 9.

38 See 'Working in analysis' on the Civil Service careers website (https://bit.ly/2RrJ0Uq).

39 Gemma, 2014, 'Chocks away – tales from Ordnance Survey's flying unit', Ordnance Survey, 11 June (https://bit.ly/3hB1ZH2).

40 A useful website when it comes to jobs in this sector is https://jobs.planningresource.co.uk/.

41 Céline Chhea, 'Lesson eight: humanitarian relief', Geographical Association (https://bit.ly/3klXou5).

42 Ibid.

43 'The plus of Erasmus+', Erasmus+, European Commission (https://bit.ly/3bYEbvp).

44 'Occupational maps', Institute for Apprenticeships and Technical Education (https://bit.ly/32tn0z5).

45 A list of geography apprenticeships can be found on the website of *The Apprenticeship Guide* (https://bit.ly/35GIE4y).

46 Jemma Smith, 2020, 'Virtual work experience', *Prospects*, June (https://bit.ly/35zUe1F).

47 'Ethical volunteering abroad', WorkingAbroad (https://bit.ly/3huyvKK).

48 AGCAS Editors, 2020, 'Geography', *Prospects*, August (https://bit.ly/3muNyYG).

49 Simon Jenkins, 2007, 'The assault on geography breeds ignorance and erodes nationhood', *Guardian*, 16 November (https://bit.ly/2FBYFxK).

50 Q&A with Kate Edwards, CEO and Principal Consultant, Royal Geographical Society (https://bit.ly/3mgmCvH).

51 'Choosing a career with geography', Royal Geographical Society (https://bit.ly/3ipdaDL).

52 Woman's Hour, 2020, 'Singer Laura Wright, Covid-19 and oestrogen, Professor Heather Viles and Covid-19 and fashion', *BBC*, 22 May (https://bbc.in/2RpQtDA).

53 Ian Cook *et al.*, 2018, 'Inviting construction: Primark, Rana Plaza and Political LEGO', *Transactions of the Institute of British Geographers* **43**(3), 477–495 (https://bit.ly/3mjLr9Y).

54 Ian Cook et al., 'How do you come up with an interesting research idea?', Video series on the Time for Geography website (https://bit.ly/3c35e8G).

55 Alan Kinder, 'The geography national curriculum', video created for the Geographical Association (https://bit.ly/3ksQoM3).

56 UCAS, 'Geography', Subject Guide 2020 (https://bit.ly/3hCvmsx).

57 Russell Group, 'Our universities' (https://bit.ly/2DZnrYh).

58 Russell Group | Informed Choices, 'Geography' (https://bit.ly/2GYTb0W).

59 Office for Students, 'Teaching' (https://bit.ly/2RlZbCS).

60 Office for Students, 'National Student Survey – guide for students' (https://bit.ly/3bYsUey).

61 For more on the 'Programme accreditation' scheme, visit the RGS-IBG website (https://bit.ly/3mi7vBL).

62 'How to write a personal statement for geography', YouTube video, The Curious Geographer (https://bit.ly/2GR7PHk).

63 'Personal statements – finding a formula', YouTube video, UCAS (https://bit.ly/3huBF0X).

64 'Come rain or shine: understanding the weather', online course from FutureLearn (https://bit.ly/2Zw3Iaj).

65 'The importance of geography in a pandemic', YouTube video, James Cheshire (https://youtu.be/VYe_4-SWIZ4).

66 'Subject benchmark statement', QAA (https://bit.ly/3ivqSFg).

67 M. Solem, N. Huynh and J. Kerski, 2019, 'Teaching geography students about careers', in *Handbook for Teaching and Learning in Geography* (Cheltenham: Edward Elgar).

68 'Careers in geography', American Association of Geographers (https://bit.ly/2FjjVc5).

69 Eeva Kemppainen and Anna Ylä-Anttila, 2015, 'Teaching media literacy & geographies of consumption', Pro Ethical Trade Finland (Eettisen kaupan puolesta ry). A copy of the guide can be downloaded from https://bit.ly/35CJHTj.

70 Steve Brace, 2020, 'Gender balance in geography', *Gender Action*, 15 January (https://bit.ly/3huliBF).

71 Danny Dorling, 2020, 'Geography and the shifting ratios of inequality – university, A levels and GCSEs in 2020', *Geography Directions*, 21 August (https://bit.ly/35zpnlz).

72 'Choosing geography at GCSE and A level', Royal Geographical Society (https://bit.ly/3mkuTi5).

73 Phil Penfold, 2019, 'Common sense tells us it's a terrible idea to build on flood plains, says Yorkshire gardening guru Alan Titchmarsh', *Yorkshire Post*, 22 December (https://bit.ly/33xlzyJ).

74 Richard Bustin, David Lambert and Sirpa Tani, 2020, 'The development of GeoCapabilities: reflections, and the spread of an idea', *International Research in Geographical and Environmental Education* **29**(3), 201–205.

75 Ibid.

76 'eConference 2020 – public lecture', YouTube video, Geographical Association (https://youtu.be/y6Kc-rSQCbU).

77 'Everyone is a geographer', Royal Scottish Geographical Society (https://bit.ly/32rDQhy).

78 Murphy (2018): see note 9.

79 Editorial, 2020, 'The Coronavirus crisis: what will the post-pandemic city look like?', *Urban Analytics and City Science* **47**(4), 547–552 (https://bit.ly/2FttGnR).

80 See note 36.

81 Daniel Raven-Ellison, 2017, 'Why all geography teachers should teach their students to be explorers', *Geographical*, 10 October (https://bit.ly/3hrzsU1).

82 Oliver Jeffers, 2020, 'An ode to living on earth', *TED Talk*, April (https://bit.ly/3ivB4O2).

83 'Young Geographer of the Year', Royal Geographical Association (https://bit.ly/3ixXuy3).

84 'Ron Cooke award', Royal Geographical Society (https://bit.ly/3mkKhuL).

85 Zersorger, 'Bad geology films', *Letterboxd* (https://bit.ly/35BPWXA).

86 Simon Brandon, 2020, 'Celebrities are helping the UK's schoolchildren learn during lockdown', *World Economic Forum COVID Action Platform*, 21 April (https://bit.ly/2ZAV9L8).

87 'Living adventurously', Podcast series (https://bit.ly/2ZDKEXD).

88 Tim Harford, '50 things that made the modern economy', BBC podcast series (https://bbc.in/35FV9h7).

89 'Ask the geographer', Royal Geographical Society podcast series (https://bit.ly/2FjtM1z).

90 'GeogPod – the GA's podcast', Geographical Association podcast series (https://bit.ly/35xFQH6).

91 'Costing the earth', BBC podcast series (https://bbc.in/3iI74Pb).

92 'The inquiry', BBC podcast series (https://bbc.in/3khEkNu).

93 'The documentary podcast', BBC podcast series (https://bbc.in/2ZAvmTw).

94 'GA branches', Geographical Association (https://bit.ly/3mjcySs).

95 'What is geography?', *National Geographic* (https://bit.ly/33qdQm5).

ABOUT THE AUTHOR

ALAN PARKINSON is currently head of geography at King's Ely Junior and has more than 25 years of teaching experience. He previously spent three years leading on secondary school curriculum development for the Geographical Association, of which he is a trustee. He has worked with schools and universities across Europe on ERASMUS projects, and he's authored resources for a wide range of organizations including Google, the BBC, Costa, the British Red Cross, the South Georgia Heritage Trust and TUI. He is the co-founder of Mission:Explore and worked on its award-winning series of children's books. He has written and edited textbooks for all stages of school geography. He worked as an Ordnance Survey GetOutside Champion between 2018 and 2020, promoting the benefits of time spent outside. Alan is a Fellow of the Royal Geographical Society and of the Royal Scottish Geographical Society, and he's also a Chartered Geographer. He is a prolific blogger, notably at http://livinggeography.blogspot.com, and he tweets using @GeoBlogs.

He is the Vice President of the Geographical Association for 2020–21 and is scheduled to be President for 2021–22.